T0312769

Praise for *Flint*

'If you like walking through the English countryside and the deep history of these isles, you will adore this vivid, personal, upbeat book about the hundred varieties of flint gleaming just under our feet. It's an archaeologist's love letter to a landscape trampled by prehistoric elephants, bears, boars, Romans, Saxons, Romanies and modern picnickers. Joanne Bourne makes the reader feel her happiness as she spots in a wood or on a chalk beach yet another shape or colour of the ancient stone that obsesses her, or as a Red Admiral butterfly curls its tongue for the salt on her arm'

Maggie Gee

'*Flint* is a beautifully written love letter to the history and mystique of a stone that has shaped human civilisation for millennia. Joanne Bourne's enchanting narrative and personal anecdotes bring to life the magic and enduring significance of flint in a way that is both educational and deeply heartfelt'

Alastair Humphreys

'Joanne Bourne writes beautifully and convincingly. I liked it very much and learned a lot'

Liz Trenow

'A unique, well-informed and enjoyable read that puts a new slant on this wondrous material from prehistory to the modern day'

Nick Card,
director of the Ness of Brodgar

FLINT

Published in 2024
by Eye Books Ltd
29A Barrow Street
Much Wenlock
Shropshire
TF13 6EN

www.eye-books.com

ISBN: 9781785634086

Cover design & layout by Nell Wood

Typeset in Vendetta OT and Brother 1816

British Library Cataloguing in Publication Data

A catalogue record for this book is available from the British Library.

FLINT

A Lithic Love Letter

JOANNE BOURNE

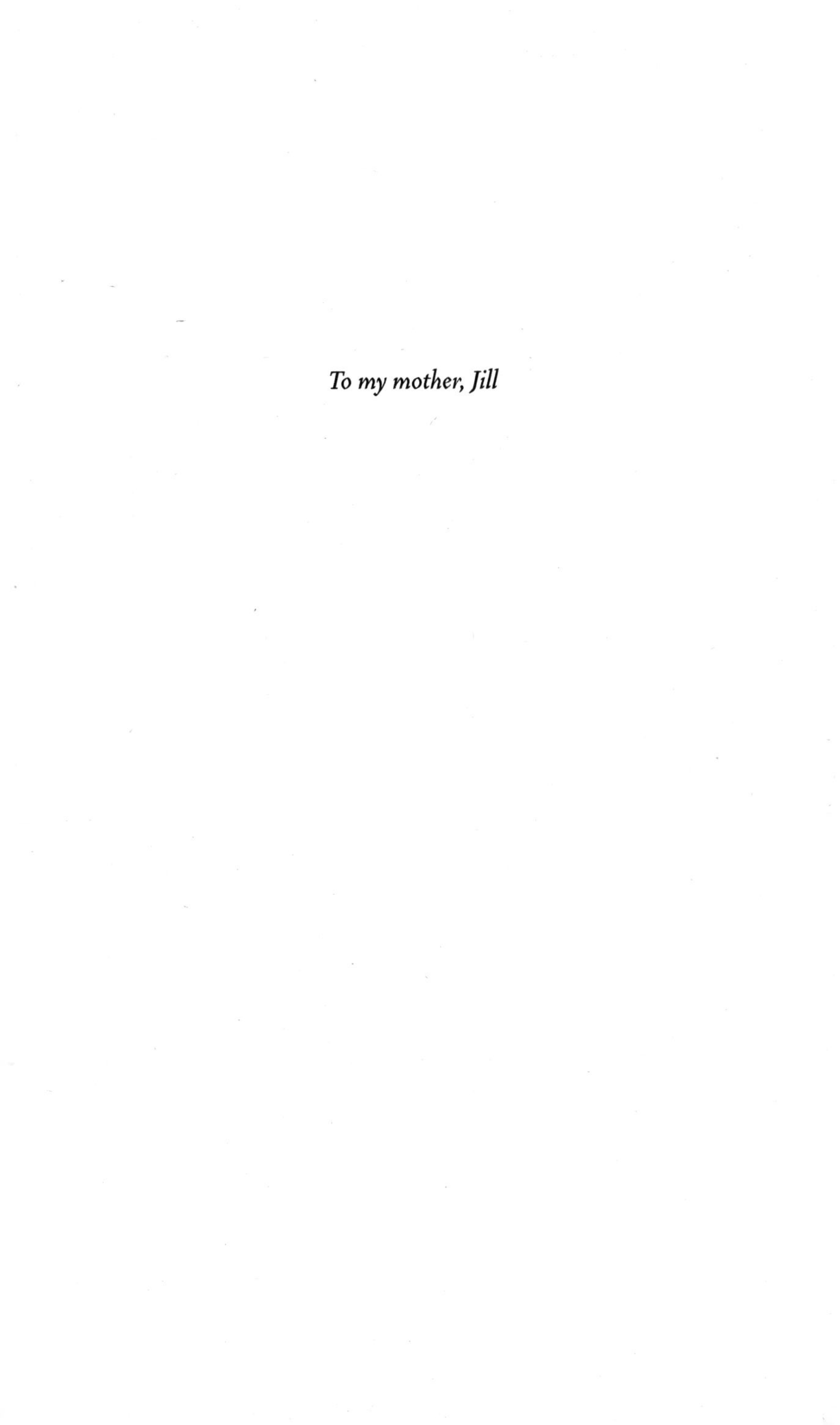

To my mother, Jill

CONTENTS

chapter one

DOG SKULL AFTERNOON

Autumn had come late to the Downs that year and the leaves hung tawny and bleached green on the trees. The day was still; a minuscule shift in the air currents brought them down around us in eddies, as we walked – Frank Beresford and I – up a flint track on a pale November morning.

Frank Beresford was a retired schools inspector from West Wickham. We had met just once, that summer, at the Lithic Studies Society in London, at an afternoon talk on the artefact-logging systems of the Portable Antiquities Scheme.

We were the kind of audience that shuffled in and nodded where necessary before fixing on the speaker or our feet. But that day the president, in a burst of embracive enthusiasm, asked that we each introduce ourselves and declare our area of interest. When my turn came I kept it brief, telling the group about my fieldwalking, the flint tools I had logged and the Neolithic settlement I had roughly mapped on the upper slope of a river terrace on the North Downs. What might have been truer to say was that I was a 54-year-old woman with a rock problem, but it was not a confessional, and anyway, we all had our issues.

Frank Beresford's, it turned out, was handaxes. Frank was writing a report on the Palaeolithic tools of the Upper Ravensbourne valley; an axe had been found locally to me by a 19th-century antiquarian, and Frank said he'd be keen to join me fieldwalking, should ever I happen to be over that way, to get a feel for the landscape.

And here, some months later, we were.

I had grown up on these Downs, on the opposite side of the valley. I still live in the village where I was born, which is mostly happenstance and suggests a continuity that's by no means true, but that is one of the great blessings of my life. From my house I can see fields and the small copses that stretch away to the wooded ridge that defines a wide horizon. I walk the Downs all the time, but not these woods and fields in my immediate view, which are farmland and mostly out of bounds.

The track we were following – a hollow way worn to mud and stone by footfall and washed deep in the middle by rain – runs at right angles from the valley floor up to the ridge, after which it doglegs down into the next valley. The fields on either side of the track have no paths across or round. A warning to trespassers on a sign close to the entrance stile, and again farther up the track, must have put the local dog walkers off. The couple of times I'd ignored the sign and walked the fields after ploughing, I'd turned up nothing for my trouble. Not even a decent nodule. I'd noted them as archaeologically sterile and geologically dull. Anyway, some places just feel wrong if you're alone. It was years since I'd walked this way, so I was happy to be exploring afresh in Frank Beresford's company.

The find spot of the axe Frank was studying was a small chalk quarry near the southern tip of the wood where we were headed. Frank was certain it came from the clay overlying the quarry and was unearthed as estate workers dug down to get to the chalk to use for fertiliser or to make lime for mortar.

The axe was now in the British Museum stores – part of its vast back-room flint collection. Frank showed me a picture on a printout. It was magnificent. Deep-toffee-coloured flint had been knapped – that is, shaped with a series of carefully aimed hammer blows from another rock or antler piece – into a round-tipped kite-shaped axe that would fit into the grip of a large man's hand.

Its date – based on others of its type – was around 420,000 years Before Present. It meant that no *sapiens* could have made it. The hand that knapped the tool belonged to *Homo heidelbergensis*, a species of human that lived 800,000 to 300,000 years ago – becoming extinct around 100,000 years before the *sapiens* lineage emerged on the African continent.

I have held tools not touched since they were discarded 5,000 years ago in the late Stone Age. I have gathered flint flakes struck by Neolithic knappers on the banks of long-dead rivers. I've found fossils worn smooth by ancient human touch. But I have never held a tool made by another kind of human. I asked Frank if he'd had the axe out of its storeroom case. He said he had. I wanted to ask what it felt like to have it in his hand, but that was at once stupid and personal. Stupid because it probably felt like nothing; personal because if it did feel like something, that was between Frank and the hominin.

The hollow way was dry underfoot. Flint – variously rounded or chipped and sharp – poked through the earth and the leaves, and our feet beat a faint rap. Beneath us, beneath the metre or so of soil, was around 300 metres of chalk, sitting in turn on the clays and the greensands of the South. Walkers – such as myself – and runners can detect the shift in geology from the resonant, hollow-sounding chalk to the dense, dull greensand through their footfall.

I wanted to know more about the alien axe. Frank said he had pondered its use. Southern England at the time the axe was made basked in a lush, swampy warmth known in geological terms as the Hoxnian interglacial. It was not an England anyone today would recognise. Among the beasts that roamed the landscape was the now-extinct straight-tusked elephant, a creature at least twice the size of today's African elephant. In parts of Europe – notably Essex, just fifty miles from where we were walking – large male elephants in their prime had been speared and their carcasses butchered using tools such as Frank's axe. Or perhaps the axe gutted a rhinoceros or a Barbary macaque. Or perhaps it smashed the scavenged bones of a lion for its marrow.

Birds sang overhead. A plane came over low, heading for Biggin Hill. It was hard to imagine a lion carcass butchered on the downland I could see from my home.

The woods thinned. We were high now and could see back across the valley to the village. I could just make out the roof of my house at the farm's edge. I pointed it out to Frank. He spoke about the river that had likely flowed in the valley between – down where the road was now. I imagined the hominins walking down to the river (stately, and in slow motion). Frank said it was almost impossible to know the form the landscape took at the time of the axe-maker's people. Before these groups colonised Britain, great glaciers from a period known as the Anglian Stage – the most severe ice age in the last two million years – stretched their fingers as far south as Hornchurch in Essex, just twenty-eight miles north of here. Then, the land we now stood on had been tundra. The extreme, rock-shattering freeze-thawing at the epoch's end turned the ground surface to warm slurry cut by fast-flowing rivers. The surface in turn refroze and cracked, then baked, then flooded. The shifting conditions turned the whole area into a great periglacial stew made of churned upper levels of the chalk, the flints that had originally lain within it, and all the soil and vegetation that had been present before the freeze. What we were walking on – the clay with all the flints in it – was dried stew. When the ice finally retreated, the area became a swamp, which warmed then sprouted tropical vegetation, providing a handsome and food-rich habitat for hominins living at the northernmost edge of the world.

This was deep human time. Not as deep as deep Earth time and nowhere near as vast as space time, but just as hard to grasp. If I held the hand of my mother, and she held the hand of hers and so on, we would create a chain of 20,000 generations before reaching our indirect ancestors who lived in this valley – a chain that would stretch twelve-and-a-half miles, or twice the entire length of Frank's and my proposed walk.

We wandered on up the track, and the woods thickened. It was time to abandon the path, and we cut left through the trees to the field's edge. From across the valley this field appeared a dull brown. I had assumed it had been ploughed and never reseeded. I was wrong. The crop – something I couldn't even recognise – had not been reaped and had gone to seed. It seemed wasteful and sad. It probably would have cost more to harvest than the crop was worth. The brittle stalks reached almost to the top of Frank's head. Between were all manner of weeds. Below our feet the ground was gravelly – flint in all colours on a gingery clay, including some small chips the toffee-gold of Frank's axe.

We skirted the crop's edge until I found a way through. It

was a worn path, made by badgers, foxes and deer, by the side of it. Not quite suitable for humans, but it was what we had. I went first. Burrs from the dead crop snagged our clothes and caught in our hair. Thistle seeds escaped as our shoulders brushed the weeds among the skeleton stems. We stepped carefully over animal scat, bending every few moments to check the flint. The path opened into a clearing some four metres wide where the earth was especially gravelly. Frank gathered samples but they were just that. No tool fragments, no waste flakes. Just old rolled flint.

We were walking on ancient stuff, a thousand times ploughed, with vegetation that was completely humanly created. And yet the path, the field and the whole rise from the valley had a profound wildness to it. I felt the soft, mothy flutter of panic. The wild, trickster god whispered, and I was scared. I swallowed it down, narrowing the gap between Frank and me for comfort. As a distraction, I tried to concentrate on the distant swish-hum of the cars on the A21. I failed.

So I tried to picture a landscape of elephants. And failed.

Then I tried to picture the axe-maker, holding a spear, following a lion from the phantom-maybe-riverbank to the tree line on the ridge. A tree line that would not exist for perhaps another 500,000 years.

I failed. It was all so long gone.

I mentally let go of the hand of my mother, and all 20,000 of my antecedents went spiralling off into the periglacial stew.

But my discomfort in the wildness had subsided and I was hungry.

Looking down, I saw a small heart-shaped fossil lying on the gravel – a pretty sea urchin of a type known as a micraster. They always feel like a gift from Earth to me. These sea urchins, or echinoderms, from the same family as starfish, lived on the beds of the shallow Cretaceous seas that covered Europe in the last days of the dinosaurs. But hunger had made me bored with time, shallow and deep, and with the fossil in my hand, I followed Frank Beresford into the copse ahead for some lunch.

We found a fallen tree that was long enough to seat us both. There was no view, just trees, but it was nice to be picnicking in the woods. Frank got a large brown-bread sandwich out of a Tupperware container. Cheese and pickle, I guessed. I took a Waitrose mince pie and a square of Gruyère from a misted takeaway box. A robin perched on a low branch and sang. I watched his little beak and red breast giving their all to the tune. I took a bite of my mince pie, and a bite of the cheese. I sipped some coffee from a flask. Frank had brought tea. We agreed, Frank and I, how good it was to eat out of doors. He'd picnicked on a hillfort earlier in the week, and the week before above a Roman site. I was a bit envious.

We talked about Christmas, and how, nice though it was, it would interrupt our stuff. Not the normal working, shopping, eating, sleeping, but stuff such as we liked doing; in my case scouting for flint, eating out-of-doors and thinking. There was never enough eating out-of-doors and thinking.

Lunch over, we packed up our things and left the copse. It was a short walk along the edge of a field boundary to the next wood, which looked like a lopsided diamond on the map and sloped from midway up the valley to the ridge. Here, in the long-gone quarry, the axe had reportedly been found.

The wood was surrounded with solid post-and-wire fencing. We scaled it – just, without losing too much dignity. Dense, dark ivy covered the ground within. Ivy and... bottles. Truth be told there was no ground visible. Just ivy, fallen leaves and a great scree of 19th-century green bottles stretching away on either side of us and ahead up a steepish slope. It was not so much the fact of the bottles themselves but the quantity that was so mind bending. I poked around the glass shingle with my trowel. There seemed to be bottles to a depth of several centimetres. A few had lettering on – in some cases complete words: ABBEY and ROMFORD and LONDON. TRUMAN & CO and, best of all, FREMLIN. Frank found one that was nearly complete, but for a chip in the rim. He inverted it and peaty black water sputtered out. I picked up a fragment with good lettering and put it in my bag for reference.

It was baffling, though. Who would dump so many bottles in a wood so far from anywhere? Frank suggested a Victorian pub might have emptied its cellar, but I doubted this. I knew the pubs and it just seemed too far to lug the stuff across fields, whether by hand or by cart.

We picked our way over the fragments and up the slope, reaching a flatter, bottle-less part of the wood.

There, running – as far as we could see – right up the woodland's centre, were two parallel earth banks, each around a metre in height and wider across at the base. Mature beech and sycamore grew, both from the banks and from the narrow ditch between. Frank thought the banks might have been a land boundary. It certainly looked too narrow to be a track – those locally had proportions this ditch lacked.

Someone had used it as a road at some point, though. For in the centre of the ditch, looking like the mechanical equivalent of a scavenged elephant carcass, was a slumped and seriously decaying white Mini van. Its front axle had been torn out and lay farther down the hollow. The detached bonnet was sinking against the far bank. Part of a door – or maybe part of the wing – was leaning on a tree. Inside the van, peeking out of the deep leaf cover on the floor, were the skeletal frames of the front seat-backs and the steering-wheel top. The white roof held, its entirety speckled by little yellow leafy kisses from the birch trees above. I looked inside again. Everything was furred with a downy rust. Outside, the wood was strewn with orange and lime sycamore leaves, fluorescing out an entire summer of sunshine.

Frank was shaking his head. Who would abandon a van in a wood? Unless it was stolen and dumped? Then I saw another partial car carcass behind a coppiced tree. Blue and twisted. Not much of it left.

Of course. This wood was just half a mile from a Traveller encampment in the valley that had been occupied throughout much of the sixties and into the mid-seventies. It was a place of caravans, bonfires and barking dogs that I remembered knowing well, but only from a distance, and hadn't thought about for years. When they all finally departed, the council had cleared and bulldozed their camp, leaving nothing other than a dark stain in the ploughsoil. The Travellers – who dealt in scrap – must have driven, towed or heaved the dead cars and trucks into the woods together with whatever other rubbish they had, and nature had done its best to eat it. A decade from now it would be gone. It was like a body farm for vehicles.

We wandered round the dead cars in wonderment. My expertise in seventies vehicles was limited; Frank's was only slightly better, but then, it was hard to identify them by rusted parts, just as it's often hard to identify an animal by a single bone.

Frank wanted to find the Victorian quarry, though we reckoned we already had; filled in with bottles from goodness-knew-where, but just to be sure, we followed a path down towards the lowest point of the wood.

'Here it is!' said Frank. He had been studying the OS map. But I couldn't see a quarry – just a mound.

'Here, it says "earthwork",' said Frank, indicating the map. It did, and it was. There was no further explanation as to what nature of earthwork, and I couldn't read it as being anything I recognised. Frank was intrigued.

I left him studying the mound, his face golden in the leaflight as if underlit by a hundred buttercups. The darkest days of the woods are in summer when the tree canopy shields the sunlight. Late autumn the woods were now light, yet somehow today they were extra light. As if the leaves were releasing all the sunlight they had absorbed during their season of life, into a final blazing aria. I wandered back up the dip of the double bank. A large sycamore to the left of me had shrugged off all its leaves; some had fallen artfully onto the stump of a dead companion. I took a photograph. As I moved around the stump for the best angle, I could see beyond to a tableau so incongruous I almost laughed. A yellow Ford Zephyr was set into a tangle of scrubby trees and vines, nose down, back end in the air. It was shored up by ramps of gingery clay, dotted weakly with flint and streaked with chalk. The dead car formed the centrepiece – like some great fascinator on the head of an earth-giant bound for a wedding – of one of the biggest live badger setts I've ever seen. It was a work of art. And unnerving.

'*Founnnd* something!' It was Frank, down the track. An exclamation that held the ghost of a shriek.

'I think...it's a skull!'

It's only afterwards that you can comb out the thoughts that must've all come at once and went something like: '... could it be a human skull/if so we'll have to phone the police/we can't phone without leaving the wood as neither of us has had a signal for about an hour/if it is a human skull I can't have it/if it is a human skull I wonder if it belonged to a gypsy/is the skull in a car?/I want it whatever it is/does that make me bad?/I think it does...'

Frank was standing near the edge of the earthwork, pointing to something about a metre away from him. 'I thought it was a piece of flint...' He let the sentence hang.

Nested in the leaves and the ivy was the skull of a small mammal. It was about the size of a badger's, without the raised crest along the top where the chewing muscles attach, but with some serious incisors. The bone was quite degraded, and it had a mossy green stain across the cranium. It was a little dog, probably quite young when it died. I wondered what it was doing so far from anywhere then realised it probably belonged to the Travellers.

'Do you want it?' I asked Frank.

'Goodness, no!' he said.

'So I can have it?' This was brilliant.

'No, please, have it. Feel free.'

He couldn't believe I wanted it as much as I couldn't believe he didn't.

I dug around in my bag for a plastic food container. I tucked it away, padded with a couple of supermarket bags, in the bottom of my rucksack.

It was starting to chill. We agreed it was time to go. We took an easier route from below the earthwork out of the opposite side of the wood. We crossed the full width of the field, cutting diagonally through the dead crop where there was a clearish path. It took us to the track near the ridge, and we straddled a broken fence and started walking down to the valley. Frank had found nothing and I was sorry for that, but he seemed pleased with his afternoon, and we chatted happily about the coming weekend. He said he'd send me the report on the axe when he'd written it up.

As for me, I was delighted. I'd found no humanly worked flint – again – but I had a fossil and a dog skull.

Then we saw it. Hanging in a hawthorn tree to the right of the track. It was impossible to miss. The most enormous brassiere I had ever seen. It was white – or had been once; now it was a yellow-grey – and showed signs of extreme wear. I recognised it as a type sold in Marks & Spencer in the eighties. It was suspended by the straps and rippled gently in the late-afternoon breeze. At that height, it would have taken some arranging. Frank stood respectfully back while I got closer. I reached up for the label by the clasp, but the size had long since been worn from the ribbon. I was disappointed, but disappointment seemed prurient. This was someone's spent foundation garment and how it ended up here was anyone's guess. I stepped away, but all the same, took a photo of the hawthorn bra in all its glory.

It was an odd thing to find and was somehow inappropriate, there at the sunken lane's edge. I felt uncomfortable, as – from his distance – did Frank.

'I wonder who it belonged to?' I said, in an effort to dissolve our mutual embarrassment.

'Must have been a goddess,' replied Frank, emphatically.

And we made our way – somewhat awkwardly – home.

chapter two

FLINT: A LOVE STORY

I am a flint addict. An obsessive. A junkie. Or something like that.

The shelves in my house, all mantelpieces, some designated floor areas and various parts of the garden are miniature sculpture parks of this wondrous biogenic rock, a form of quartz that's made from the remains of once-living creatures. It is sorted by shade, shape, size, type and whim. There are bowls of it, glass jars, display cases and wooden trays. I file it under my bed in buff-coloured archive boxes; I keep it in my shed. And in my mother's shed. And once in a friend's loft. I'm as happy to display it as any piece of art, and I have the satisfaction of having recognised its specialness, picked it from a ploughed field and lugged it – often several miles – home in a bag on my back.

I've been in awe of flint as long as I can remember. It was all around me on the Downs, growing up; so commonplace that it was used as a building material for garden walls and rockeries, to edge drives and weight dustbin lids. More exotically it was the stone chosen to build pubs and churches, the foundations of two Roman villas and a bathhouse, and several castles. It was the local stone; for centuries it was all anyone had to immediate hand, and it had only to be dug out of the ground.

My first experience of flint was a wall at bottom of my lane. Around two feet high and topped with brick, it fronted the raised lawns of a row of cottages, curving around the end cottage to the side road. My mother would stop to chat to neighbours on this corner; I hung on to the chrome bar of my brother's pram with one hand and traced the faces of the struck flint with the other.

The smooth of it!

Purply black with a thick white rind, chocked with porridgy lime mortar. Milky grey and speckled. Some with tiny sparkling crystals or golden orange iron flecks or banding.

Tiny red spider mites milled across their knapped faces in the sunshine. Small ferny plants grew from the mortar cracks. I was a very small child; when no one was looking I could get really close to the flint and touch my nose against it.

I knew then, and know more now, that flint is a magical stone. It is commonly cited as being second only in hardness to a diamond, can be struck to make a spark that will ignite a fire and has been part of human development for around 3 million years. Mined, worked, traded, used as a weapon to hunt and in war, cached or hung as protection against thunderbolts and fairies, kept as a talisman, ground to fragments and set in the base of Roman mortar bowls, used by the ancient Egyptians to make extraordinary hooped bracelets and as a circumcision tool in the Bible, it made us, as much as we made things with it.

Once, a Balkan archaeologist, working in a Dalmatian valley, tried to tell me about flint. I told him I not only knew what it was, but came from a village where the houses were built of it. He looked doubtful and said that surely they would attract lightning and all be burnt to the ground. Then he went off, scoffing. I choose to believe this said more about his reverence for flint than his misunderstanding of physics. And anyway, if all you had known of flint were some tiny chipped arrowheads and scrapers, how could you conceive of a house built out of the stuff? Nor actually, what natural flint looks like?

Because flint – in its original form – looks like all sorts

of things. A dragon. A child's drawing of a fish. A small poo. A stylised bird. A ready-to-bake croissant before it's arranged in a crescent on a baking tray. A tennis ball. A sausage. It has a white-to-beige skin and is spiked and curved, globular and finger like. Its patterns repeat but, like snowflakes, no two pieces are ever the same.

My primary-school playground ran alongside a cul-de-sac of chalet bungalows whose immaculate rose beds were separated from our tarmac by a galvanised chain-link fence. Twice I wriggled my arm through the links for treasure from the dug-over soil: once for a small fossil sea urchin, and again, for the forked claw of a dinosaur. Or so I believed the latter to be for many years, and though it wasn't so, what I had found – and the story of how it came into being – is equally strange.

Some 144 million years ago, the world was warmer. The poles were forested and global sea levels higher than at any time before or since. Much of what had once been land was washed with shallow seas that rose and fell, covering parts of Australia, Africa, Canada, South America, Russia, India and Western Europe.

All kinds of creatures flourished in these waters – and died. A constant 'snowfall' of tiny organisms, whose shells were made of the mineral calcite, drifted to the seabed, where they settled and consolidated – over time – to great depths, to form chalk. If you were to scrape some crumbs from the White Cliffs of Dover and put them under an electron microscope, you'd see tiny, spoked ovals (shaped like the plastic baskets that hold the hot towels in Indian restaurants) called coccoliths, shed from micro-organisms called coccolithophores.

Into these powdery, micro-skeletal seabed drifts, flint was born. Geologists are fairly certain they know how it happened, though, unlike with chalk, they can't be sure. What they believe is that silica-based plankton and the skeletons of sea sponges dissolved after their deaths in the seas, making the waters intensely silica rich. Over time, this water seeped into the twisty burrows of bed-dwelling creatures, trickled into fissures in the hardening sea floor and filled the empty shells of larger deceased sea creatures. There, the silica settled out of the water and slowly hardened – taking on the exact shape of the burrows, fissures or creature remains. Sometimes it collected around an object's surface like frost on a leaf.

Or an art project of the great divine.

There are places on the Downs where the topsoil has no depth at all and ploughing periodically scrapes the fields down to the chalk. Here, machines can toss up whole flints – lumpy masses known as nodules – their shapes the perfect positive of the negative they formed in. Pure, without cracks, they ring if you tap them with a fingernail, especially the long ones. These are the gallery flints – the Barbara Hepworths and Henry Moores. The Mirós and Picassos.

The deep flint – dug more than a metre down – is often

much larger here and has the brittleness of just-crunched butterscotch. It rarely comes out of the chalk intact. I rescued a barrowload of nodules from a neighbour's extension foundations once, almost perfect, covered in clumps of wet, sticky, yellow-white chalk, like the stuff on a newborn baby's head. The sheared edges were so sharp I cut both hands without even realising until bright-red blood started pouring down the barrow handles. It made me think of the young flint miners descending into the deep-shaft flint mines in Neolithic times. The mines – circular shafts sunk as deep as thirteen metres to access bands of flint for tool working – were dug with antler picks and accessed by wooden ladders. At Grimes Graves in Norfolk, one of 400-plus flint mines on the site's rolling heathland is open to the public and is a marvel of prehistoric engineering. The original miners, picking out nodules in galleries at the base, would have known that flint could bite. Maybe they had leather wraps for their hands.

I hope they did. The cuts on my fingers from my neighbour's nodules were fine sliced but they stung to buggery.

The outer surface of flint – its cortex – is a little like the rind of oranges, in that it's sometimes thick and pithy and sometimes barely there. This skin of flint – for that's what it is – is formed from fine-grained opaline silica. This is not opal in the sense of the fiery rainbow stones of Coober Pedy in Australia (though that also has a similar biogenic origin), but it manifests in shades of yellow, blue, pink and purest white. This interface between 'flint' and 'not flint', once exposed, weathers to a million buffs, ecrus, beiges and even blacks. And there begins a whole new appeal, for the pitting, scarring, whorling and inclusions – like mini coral segments or seashell prints – is a wonder in itself.

Most of the flint we see that freckles ploughed fields and garden beds is broken, splintered by landslip, freeze-thaw, weathering and ploughing. My dinosaur claw, found in the cul-de-sac rose bed, was a cast of the twin end-points of a tiny shrimp-like sea creature's home, snapped from the main burrow possibly millions of years ago. Fossil dinosaur bones do exist in the chalk, but they are rare as hens' teeth, for dinosaurs lived on land and all that is flint and chalk was once water. Plesiosaur remains are more common (though still hugely rare), as these prehistoric creatures at least swam in the waters that gave rise to the chalk.

Sometimes, the flints break to reveal quartz crystals inside. As a child I had a small collection of such treasures. Once,

when builders were landscaping the church grounds behind our house, I found a large piece of flint with black-coloured crystal and recognised it as a special treasure. To the back of the grounds was a huge, weedy earth mound about three metres high that we used to play on. It must have been left over from an early sixties church hall addition – probably the spoil from the foundations. It made a good castle. One side of the mound had a hole in it – I think some kids had made it with digging sticks – and I thought this a good place to hide my find.

A week later, disaster struck. I came home from school to find the builders hacking at the mound, presumably under orders to level it. They had already razed it by a good metre. The hole – and my treasure – had gone, and our castle mound was vanishing fast. The church lay on a slope and our house on the rise above it. I stood in the garden looking down on the idiot builders through hot tears, and threw clods of earth over the fence at them. I'm a bad throw and all fell short, though they ducked, swore and yelled 'Oi!' for about a minute before I ran.

I still think about that black flint. I now know it to be rare because I have never found its like again. It's a small comfort to know that the flint with its crystal will outlive the builders. It will outlive me, too, plus all my relations and possibly their descendants. I just wish, for my time on Earth, that the black crystal flint could have lived alongside me.

Fast forward to March 2015: the day of the solar eclipse. Not a totality but 83 percent over southern England. I had arranged to work from home so I could watch it, but dawn broke and the cloud was low. It didn't look promising – on such days it so rarely did. At the scheduled time I wandered down into to my parents' garden – a field really, that ran alongside a meadow of horses. I was equipped with eclipse glasses, a camera and tripod and a piece of welder's glass for the lens. Just in case.

I found my dad at work with a mattock, hacking away at a long strip of uneven, grassless earth where brambles had been cleared to ground level a month or so previously. He resembled a bas-relief frieze character on a medieval *memento mori*, dressed in jeans and a checked shirt. He had late-stage Parkinson's and preferred not to eat or even drink much, yet he swung that mattock at the flinty earth and blackberry roots with more grace and power than any twenty-year-old road worker; a lifetime of muscle memory dialled up for the job.

'All right, Dad?'

He grunted at me in acknowlegement and kept mattocking.

I bit my lip.

A gloom was manifestly there now. The eclipse was upon us and while we were not seeing it directly, we were experiencing its effects. Small birds flitted around the hawthorns, then were still. Chirruping ceased.

My dad carried on working in the greasy light, worrying at the same spot as if resolved to hack his way to Australia. Then he tossed the mattock aside, knelt down jerkily and began clawing at something in the ground. I went over and looked into the hole. The thing inside was domed and bone-like, a dull cream, and before I could have any more concerned thoughts he prised it out with his fingers and held it high.

It was a massive flint nodule, almost spherical but, like Earth, squished at the poles. Not complete, there was some chipping on one flatter side revealing a grey, lightly patinated interior. It was crumbed in damp soil and was about the size of a toddler's head.

'That's all right, isn't it!' said my dad, pleased. 'Thought I'd found a body, though.'

Momentarily, so had I.

'Here...' he passed it to me. 'I expect you want it?'

'It's brilliant, Dad.' I took it from him. 'I've never seen anything like it.' This was true. 'Let's go up and wash it and have a cup of tea?'

My father tottered unsteadily before me, along the garden path, then into the kitchen, trailing damp sods of earth from the turn-ups of his jeans. I managed to grab a few and throw them into the yard before my mother saw.

I ran the nodule under the kitchen tap. An iron stain swirled from the top, more visible as I patted the surface dry with a towel. It really was a magnificent specimen; more than twice as large as any of the flint spheres I'd collected over the years. Later my mother found me a round wooden stand she'd had in a cupboard. It fitted perfectly and made it look like an antiquarian's prize.

My dad died the following September. His eclipse nodule was the last of a lifetime of things he dug out of the earth for me and is the most precious of all my pieces of flint. Its pocked cortex looked just like the surface of the moon, whose rare slide across the face of the sun we had failed to see on that day. It is around eighty million years old. My dad was eighty.

The flint won't help as I lose people. But it does give me perspective. This relict from a world so long ago helps me to reflect on our firefly lives. If flint could see, we individuals would move so fast we wouldn't register. Flint is like lithified time. It formed in the waters as pterosaurs wheeled overhead and plesiosaurs swam; it formed as a million things were born and died uncelebrated.

I love that we're here to wonder at it. I love that I find it beautiful. I love what I can read into it and what I can't. I love when it's just me and the flint and the thinking about it.

When I read those features that ask 'When and where were you happiest?', I know, if questioned, what I'd reply: alone, on a spring/autumn/winter morning in a sunlit Downland field. Or perhaps: at eight years old, wriggling my hand through a chain-link fence to reach for my first-ever fossil made of flint.

chapter three

JOHN LUBBOCK AND ME

'What we[...]see depends mainly on what we look for. When we turn our eyes to the sky, it is in most cases merely to see whether it is likely to rain. In the same field the farmer will notice the crop, geologists the fossils, botanists the flowers, artists the colouring, sportsmen the cover for game. Though we may all look at the same things, it does not at all follow that we should see them'

John Lubbock, 1st Baron Avebury,
in *The Beauties of Nature*, 1892

John Lubbock is tall and rangy. With the beard, he looks a bit like an ageing hipster. He has ants in his pants. Figuratively in that he is a fizzing ball of energy, and literally because in 1882 he wrote the book *Ants, Bees and Wasps*. He probably has a few specimens under current study in boxes in his trouser pockets. After all, he had a pet wasp, which he brought back with him from a trip to Spain and wrote about in his journal. But that's his story. This is mine.

John Lubbock is my walking companion. He wanders with me over the woods and fields, and sometimes he brings his dog, a big black poodle called Van. Van is extremely smart (John Lubbock once did an experiment using flash cards to teach Van to read) and gets bored with our walks. John jokes that he'd rather be at home with the newspaper, and I believe that to be true. Van tolerates me; I'm grateful he's there at all.

John Lubbock is, of course, dead. But being the man that he was, his reach extends his corporeal self. Being dead has some advantages as he has a little more time now and can absolutely choose where he goes and with whom. I'm hugely flattered.

He was born in 1834 and achieved more than many men in several lifetimes. He campaigned for the Bank Holiday – once known as 'St Lubbock's Days' – and I have often walked to his grave or memorial (not one and the same, though they're close) on those precious days of freedom and laid a feather or flower there for him. 'Why bother?' says John, 'For here I am!'

When I set out on a walk, there's a point on the path, after it leaves the open fields of Cudham North and just into the woods, where John Lubbock 'clicks in'. It's the only way to describe it. He's there waiting for me, on the borders of his estate, almost like I phoned ahead, though John has no phone. We cut up along a small path through the coppiced area, cross the badger sett below the big lynchet, scouting for any nice flints the badgers may have unearthed, and then we're away, off on another Downs adventure. For it is always an adventure and no two walks are ever the same.

Today, a Saturday in early May, we are headed for Cuckoo Wood, then on to the dene hole and back to John Lubbock's house, via the fields near Mill Hill, where the Stone Age people lived. John went all over Europe looking at prehistoric artefacts and collected many from the Downs – mostly they were brought to him by estate workers and villagers digging the valley gravel pits. In 1865 he published *Pre-historic Times as Illustrated by Ancient Remains, and the Manners and Customs of Modern Savage*, for which we forgive the title and remember that this was one of the most seminal books of prehistoric archaeology ever published. In it, he coins the terms 'Palaeolithic' for Old Stone Age and 'Neolithic' for New Stone Age. For that alone I revere him, but there was so much more besides.

He has never said whether or not he knew that there was a Neolithic settlement on his own land, but I know he revels in it. Some of the flint tool finds I make are undoubtedly mine, but some are John Lubbock's. How could it be any other way? My finds rate is impossibly high.

The green is intense on these fresh-leaved trees as the warm morning draws out the summer in everything and I can smell the caramel-vanilla scent of sun on skin. The woods are alive with a million invisible events and, beside me, John flickers, there-not-there, as the sunshine shimmies through branches.

We ascend the long, sloping path that traverses the side of a wooded hill. To the left, on the downside, I can see lumps of shiny ceramic tile; the corner of a washbasin and something that is unmistakably the rim of a WC poking out of the earth. Together with some broken bricks and quarry tiles. I reasoned this away for years until a friend who knew the woodland ranger told me the source: they were the remains of bombed houses from the Second World War, brought in trucks and scattered in the woods and across wasteland in local villages. No attempt at landscaping was made here; they just shovelled it down the bank beside the path. The same happened on the grassy mound near my local station, which has been planted with trees to mask the rubble. What else were they to do with all the detritus? There had been so much damage as German planes attempted to strike the airport at Biggin Hill.

A path bearing off through the bomb rubble leads steeply down to the valley floor and the 'enclave of the Roman snail'. These large snails are not native to the Downs, being the pale-shelled, edible variety found across Europe. The story goes that these rogue tribes are the ancestors of bold escapees from Roman settlements, whose people brought the snails to England as food, post conquest. I have long loved this romantic notion, and told the story many times but, hereabouts at least, it's not true. It was John Lubbock's family who brought them to the Downs. He laughs at this. But he doesn't know how long I spent poring over maps looking for traces of the vanished villa that might have been the Romans' home.

We bypass the path to the enclave of the snails and turn up the hill to Cuckoo Wood. The bluebells grow thinly at first, discrete explosions of purple among the green either side of the steep path. Then they start to thicken. Their intensity builds, and when we reach the plateau at the top of the hill it's all low-boughed, lime-leaved chestnut trees and fat bees, the fierce flowers pushing through brown earth, leaf litter, patinated flint gravel and last year's black chestnut shells. I follow a fox path to the densest point of the purple and sit with my back to a tree, drinking it all in. I could die happy here. Woodpeckers rap. Birdsong blurs. The scent is lovely, and I want to keep it with me for ever, but it lasts just as long as each breath, and when we leave, we will not be able to conjure it up; only remember that it was good.

John Lubbock looks up. A plane is heading for Biggin Hill, loud against the hum of several others. He says nothing.

A large bumblebee flies past us with intent, on its way in a timely manner to somewhere definite. A bee with a plan. Gnats doodle spirals in the sunbeams, but when you watch – really watch – they have attitude and purpose.

I am always delighted by the agency of insects. They still debate this, the scientists, though for John Lubbock it was never in doubt. He wrote: '...when we consider the habits of ants, their social organisation, their large communities, and elaborate habitations; their roadways, their possession of domestic animals, and even, in some cases, of slaves, it must be admitted that they have a fair claim to rank next to Man in the scale of intelligence'.

He performed hundreds of experiments with phenomenal rigour, even inventing observational equipment still in use today. Included in his findings were the facts that bees are not susceptible to the sound of the violin (nor a tuning fork, nor his hollering at their hives); ants don't just work, they play; and wasps can be trained to feed from your hand.

I unscrew the stopper from my flask and drink some coffee. Van is sniffing out something on the wood's edge. Across the flowers, John is scrutinising bark. Sunshine has found me through the chestnut leaves. The bluebells are bending under the weight of bees gathering pollen. I love this. I never want it to end.

A family with a dog walk by along the path, ten or so

metres away. They see me with my camera and wave. The dog pricks up her ears, looking in the direction of Van, who raises his head momentarily and stares. The dog lowers her eyes, and her tail, and trots on.

I gather my stuff and we move on, too, down the bridle-path, across the lane and upwards on a rutted track. We – that is, I – run under the fizzing pylon wires. I can never run fast enough, for they terrify me. Sometimes it takes me ten minutes to find the courage. John strolls, Van at his heels, whistling. My dad told me a story of a dead cow he found in a field, killed by electricity arcing down from the pylon. I don't know if it was true. How could he know the pylon did it? He said it happened in a time of intense moisture. He was an inveterate liar.

But I asked the hive mind, just to see how many cows got struck by pylons every year. None, it seems. It turns out, though, that power lines disrupt cows' magnetic compass. Birds have magnetite in their beaks so they always know which way is north, but cows? A relic from a migrational past, maybe?

The path beneath the pylons is the only way we can reach the dene hole, which lies on the edge of the wood just a few metres from a field, at a high point on the Downs.

When I was younger, the dene hole was bounded by a barbed-wire fence and was filled with rusting things. I remember a broken bike. Some twenty years ago it was cleaned out, and a metal grille fitted right across the top with a trapdoor and padlock. For safety. Solid metal railings were set in concrete. Two-bar, horizontal, painted green. The grille made me feel sad. I had fantasised many times about bringing a rope and abseiling down and exploring. But safety first, I suppose. And anyway, I have no upper-body strength.

I lean over the railings. What I am looking at is a circular pit around three metres across and eight metres deep, dug into the chalk. The top is slightly flared but the shaft is essentially vertical. Three small tunnels lead off at the base. While I can't see, I know they don't extend far. The Kent Underground Research Group investigated these strange 'Dane holes' (where the Vikings were said to have hidden), or 'Druid's lairs' (where Iron Age priests 'ritually' stored grain), and found them to be mines for obtaining chalk that was used to neutralise the acidic clay topsoil of neighbouring fields. While not as romantic as Saxon caches or Druid pits, many dene holes are old, dug in medieval times and sometimes before the Romans came. They can be dated by their pick marks – some have been dug with antler picks, suggesting they are prehistoric. Ours was of the type known as a chalkwell, dug from the 17th century to modern times.

Chalk pits considered 'true' dene holes, meaning they follow a specific pattern of a metre-wide shaft sunk up to twenty-four metres, are a local phenomenon, found only in Essex and Kent. The similarity suggests professional work. I imagine teams of itinerant hole-diggers travelling the chalklands, mining the deep, pure – described by old farmers as

'Tat' – chalk in exchange for payment. Was it a seasonal job? Was it family work, a trade passed down through the generations? Did they come knocking – the hole-diggers – offering their remedial services to farmers who were suffering from acid soil?

Once, on a Dalmatian Island in the Adriatic, I saw a paper sign in a plastic pochette, nailed to a telegraph pole. It read:

> BUŠIM BUNARE DOBRE VODE
> DO 50 M
> INFORMACIJE I PREDBILJEŽBE
> 098 786979

My companion told me it meant 'Good wells dug, to a depth of 50m', so holes – in this case wells – are dug by professionals even today.

I lean on the rails, summoning up my 17th-century pit crew. I give them a cart. Coils of hemp rope. Large wooden poles and wooden buckets. I give them a donkey. Is that even possible? I give them picks and shovels. I dress them in shades of ecru. I wonder about their boots. How would I ever know about their boots?

There are two further dene holes in these woods. That are known of. 'Beware small hollows' goes the official advice, and I do, especially after heavy rain. When the holes were finished and all the chalk extracted, it was common practice

to throw bushes, branches and even the felled trunks of trees down the shaft. Clumped and wedged often halfway down the shaft, the dead flora became covered in back-filled topsoil. When eventually the flora rotted — and it could take many, even hundreds, of years — the plug of soil at the top would collapse into the void, as happened two decades ago in the lane where I live. 'We did tell the builders,' said an elderly neighbour, 'but they wouldn't listen,' as the row of neo-Georgian, barely detached houses went up on a scrap of farmland at the top of the hill. It was some years before the front garden gave way into the chalk pit, which was investigated and reported on by the Kent Underground Research Group before being pumped full of concrete.

From the dene hole we walk down through woodland paths to the farm in the valley. The farm is old, and we walk close to its high, brick-and-flint wall along the short section of lane. We reach the road that follows the valley and cross it, walking on to a track between fields. To the left of us, the barley grows neat and fresh. To the right, a meadow not ploughed for decades. This south-facing rise was home to people 5,000 years ago and maybe much longer. I know this, because I have found several of their flint tools, several hundred flint flakes, or debitage, from the making of those tools, plus two small beads and a toe bone. I'm not totally sure about the latter, it could be part of a sheep. But it certainly looks like the mineralised bone of a human toe (middle phalanx), and so long as I declare the doubt to myself, I don't feel bad about calling it that.

There is no fieldwalking today as the crop is high. But we look at the field's edge. Flint of all kinds is lodged in the earth beneath the vigorous stalks. There are fossil sponges, shattered nodules and tabular fragments. I pick up a small sponge. It is shaped like a hot-air balloon, with small holes in the widest part. I pick up another piece of flint that looks like a finger. It is also a fossil sponge.

The barley ripples, rendering the breeze visible as its touch twists the green stalks and turns them silver in the sun. It's warm now. We follow the path northwards to another crossroads – this time just tracks, as all roads were – and carry on across a meadow to the high point on the hill.

'Look,' says John Lubbock. 'Look, the barley. Look, the shiny white patinated flint on the path. Look, the black glossy scuttling beetle.' How we love to look, John Lubbock, Van and me. Overhead, skylarks. We look up. They are loud and just above us, but we can never see them.

I don't often go to the top of the hill. One imagines tops of hills to be open to the world. Places of breezes and vistas; somewhere to sit, think and get perspective. This is not that kind of hill. First, it's thickly wooded, with small paths winding through dense undergrowth. Second, the flora is different, with Scots pine and ferns making up much of its cover. It is not of the chalk. It is alien. Exotic.

And it is all down to geology. The hill is a huge mound of sand, dumped upon the chalk when the land was a great braided river. You can see the river's like in geography textbooks, the land between the channels annotated as 'gravel banks'. The sand varies in colour. Mostly it's grey but in some places it's a deep orange – visible where a steep bank is eroding on the eastern side. Small oval pebbles of flint – almost impossibly black and shiny – spill out from the sand and wash down from the hill over the clay and chalk. They are part of this time – the Paleogene, 59 million years ago. These pebbles are common around here – and I have found them in the roots of upturned trees all over the woods and even in my garden, showing they're widespread locally. They are so regular and glossy, almost like tumbled stones from a machine. Geologists declare them to have been 'pounded together under storm conditions', doubtless in some tumultuous and enduring climatic event, and this gives me a new respect for the tiny things, now at rest in their sandy bed.

We pick our way to the top of the hill. It's not easy. Brambles snag my hair. The ferns are shoulder height and promising paths fade to nothing. When we reach the top, it's hard, even, to know we've arrived. Persisting, we walk along the ridge until we find what we've come for. A gap in the trees through which we can see – to the north around eighteen miles away – the future. It shimmers in the sunshine like a silvery mirage, though it's all too real. It is, on the eastern side, the glittery glass towers of London's Docklands, merging with the western wonders of the Shard and the Millennium Wheel. We stand there, the three of us, side by side. John Lubbock in sensible tweeds and stout boots; Van in his old leather collar; me in a pair of fraying Converse trainers and split jeans. There are small twigs in my hair. My left arm is scratched. Van is covered in burrs and has a dead thistle-head tangled in his tail. Things smell green. I feel ancient. I feel of this world, of the chalk, not that glassy, shiny world I can see on the horizon. I feel of mud and trees and sweat and blood. Yet tomorrow I shall take the train to a job in an office in one of the glass towers. I want John Lubbock to see it. Where I work. Where I have to go so that I can be here.

John Lubbock is unfazed. 'I know,' he says. And I know he knows. From the age of fifteen he travelled with his father from Kent to London to work in the family bank, first by carriage, then later by train. It was an arduous journey, as the local station would not exist for another nineteen years. He was constantly occupied – with banking and then parliamentary duties. For twenty years he served as a Liberal MP, ten as the member for Maidstone in Kent, and ten for the

University of London, before he was elevated to the House of Lords. He fought to improve education and for social reform. His interests included travelling, entertaining, ant-keeping, plant and insect study, archaeological artefact collecting and theorising. He was an inventor. He wrote books and presented scientific papers. He was twice married and had eleven children. He was Charles Darwin's friend. He was busy but he seized every bit of life in the cracks.

For all his privilege, I'm humbled by him. We even owe John Lubbock the prehistoric stone circle of Avebury in Wiltshire – the largest in Britain – which he purchased in 1871 to save it from destruction when local farmers began to break up the stones. In 1873 he bought nearby Silbury Hill, the largest artificial prehistoric mound in Europe. Aware of the threat to the archaeological and architectural heritage of Great Britain and Ireland, he worked for ten years to get the Ancient Monuments Protection Act, 1882 passed by parliament.

John Lubbock gave us a past, as well as a future, and when he became a peer in 1900 he took the title Baron Avebury.

I look at that quicksilver horizon and I want him to know that I take nothing for granted. This walk costs nothing, in that I had the great good fortune to be born on the Downs and can step out of my door in old shoes and vanish into it. In John Lubbock's time, this walk – the unhurried exploration of this gentle wild – would not have been possible for anyone but a fraction of society. But John Lubbock made a difference to the common people, and during his thirty years in parliament he worked to lift the unremitting labour of clerks and shop workers by passing acts that granted days off, likened – in terms of work – to Good Friday or Christmas Day. He also petitioned to cut hours for city clerks and shop workers from one hundred a week, which was common practice, to no more than seventy-four, and later to grant a day of early closing.

'...personally I have always expected that the first Monday in August coming as it does in the glory of summer would eventually become the most popular holiday of the year,' he wrote.

He gave ordinary people the gift of time. A very precious thing.

We make our way down from the hilltop and take a track that skirts the village, then turn on the path to the valley and the road that runs along it, crossing it to the gates of John Lubbock's house. There are cars in the car park, and an ice-cream van with a small queue. Families are walking towards the café in the old kitchen garden. Others are sitting on the sloping lawn in the sunshine.

The house – his home for almost all of his life – is gone. Lost to a fire in 1967, and I regret that I never knew it. The story is not uncommon but it's a sad one. Of the eleven children John Lubbock had with his two wives, ten survived into adulthood. John died in 1913 and was succeeded by his eldest son, John, who worked in the family bank and continued to live in the family home. It remained in the family until 1938, when it was sold to Kent County Council and the estate was included in London's Green Belt. It became a nurses' training college in 1948, then accommodation for art students at Ravensbourne College. It burnt to the ground (owing to an electrical fault) on August Bank Holiday, in an irony that does not improve with the telling.

John Lubbock's house and I overlapped on Earth by six years. I probably saw the smoke from the fire from my house as it rose above the tree line, but I have no memory of it. 'Everything dies,' John says, but I feel wistful. We approach the terraced garden and Van bounds towards the flattened lawn where the house once was, and vanishes. He's probably inside now, enjoying a bowl of water and a treat from the kitchen.

There is a bench in the shade of a copper beech, and we sit on it. My legs ache and one of my socks is rubbing inside my shoe. John Lubbock says nothing but is serene. I wonder if he's tired. Behind us is an ornamental garden, planted on the no-longer tennis courts. A toddler is running after a collared dove, which, despite the mild peril, is hoping for picnic crumbs. We look down to the no-longer house. Small boys are playing football in the afternoon sunshine. A couple walk by us and nod a greeting. Their greeting is only to me. Where John Lubbock sits there is only a tangle of light and shade cast from the beech tree above.

I pull a plaster from my bag for my foot, make it good, adjust my rucksack and start for home, up and across the corner of the golf course. Can I, through wanting and sheer force of will, keep him here? Ask any child with an imaginary friend how real he is. John Lubbock is as real as Australia to me. And all the things I know of but have never seen. I have had another lovely day and I treasure every one, while wondering if it will be the last. I reach the wood's edge and look back to the bench. He's there again – of course he is – looking across the distance, his hand raised in farewell.

It's high summer. I have charge of my seven-year-old nephew for the morning. I've bribed him to come on a walk with the lure of KitKats and a flask of hot orange squash. It's what he asked for. I have also told him we might find fossilised shark poo.

We cross the meadow using the long, diagonal track. Above us there's a lark. The lark is loud. Insistent, even.

'Do you know what that is?' I ask, pointing up into the blue.

My nephew looks up at me. 'Annoying?'

He is guileless and it's true.

We head for the stile. The path through the woods looks dark, both by contrast to the day's sunshine and because the leaves are at their fullest.

'Do you always go walking on your own – in the woods?' asks my nephew as we approach. He says it like perhaps I shouldn't. Or maybe that's what his mother says.

'On my own? Pffffh – I'm never on my own,' I say. 'I go walking with a man who used to live here a long time ago. He's called John Lubbock.'

'John Lubbock...' says my nephew, trying the name out for size. We straddle the stile and step onto the path into the woods. A large crow wanders up ahead, and a piece of flint on the path's edge catches my eye. I pause and pick it out with a fingernail and twig.

'You know John Lubbock?' says my nephew.

'Yes?' I say. I've nearly got the flint out.

'I can see him!' He nods at me emphatically and looks back across the stile. I laugh and toss the flint aside. Natural break. We stride on

'I can see him,' he says again, looking over his shoulder as we're walking. He looks serious and he knows I'm not sure about what he's saying.

'He's walking behind you!' My nephew is still nodding. He wants me to believe him. 'He's tall...and he's got a black dog!'

I turn. There's nothing on the path but the crow, pecking, looking, pecking...

I don't know what has just happened. All I know is that I feel incredibly distressed. I reach in my pocket for my sunglasses and put them on so that none of them – not my nephew, not Van and not John Lubbock – can see me cry. I cry for the possibility that he is not just my imaginary friend, all the way to Cuckoo Wood.

chapter four

HOW TO IDENTIFY HUMANLY WORKED FLINT

or

Some of my Best Friends are Knappers

Every student of archaeology is taught to identify flint that has been deliberately struck from a nodule by a human hand.

While the flints vary in size from that of a ladybird to a brick, and look – certainly in the context of a field – like the several hundred thousand other pieces lying on the surface, all have a set of physical characteristics that are instantly recognisable when you know.

First is a flat, sometimes slanted section at one end, or the 'top'. This can be large or small and is called the striking platform. (Which itself is prepared by the human hand by knocking off a chunk from a nodule.)

To the centre-front of the striking platform is the point of percussion – the precise point of the strike made by the hammerstone.

Below, on the main face of the flint, is a tiny concave chip: this is the bulbar scar, which overlays the bulb of percussion – the elegant, bulbous curve that tapers down the length of the flake, the curve repeating in visible, smiling ripples.

Back to the top: either side of the point of percussion are whiskers – small straight cracks looking just like whiskers extending from the tiny nose of a mouse.

Down again: little fissures run vertical to the ripples – these are stress splits in the fabric of the flint.

There's more, from minute, punched-out serrations along a tool edge (made by pressing the edge to make a saw-like blade with a wood or antler point) to hinge fractures – smooth but useless curves opposite the striking end (that look like the rim of a teacup) caused by a knapper's mis-strike or flaw in the flint. But this is not an instruction manual. It is enough to say that while every struck flake is different, it is also the same, whether a waste flake or finished tool. A vehicle is a vehicle whether a golf buggy or a Porsche.

How and why flint does what it does starts at a microscopic level. All minerals have a micro-structure – a three-dimensional lattice best imagined as a kids' science toy, with balls joined by sticks. Each mineral has its own lattice style, with corridors (the sticks) wide in some places, tight in others. Flint is not like other minerals in its structure, because its lattice is neat and

Lattice matters. It's the widening of planes in lattice that gives most minerals a thing called cleavage: where they are wide apart, their bond is weak, and it's here – on their planes of separation – that they'll break most naturally. Flint, with its even lattice, has no plane of separation and therefore no cleavage, so it will break in any direction. Science, but also slightly magic.

Besides cleavage (or non-cleavage), all minerals have fracture – that is, when they break, they break in certain ways. Flint is both 'splintery' and 'conchoidal'. Generally, natural breaks are splintery: the flint falls from a cliff and shatters; it is put under stress by frost or heat and its tiny but ever-present water content causes bits to break off; it gets rolled and smashed by earth movement or it gets mercilessly beaten by the blades of a mechanical plough. Intentional breaks are generally conchoidal: the swift, targeted strike creates a small 'shell-like' convex or concave shape on the broken body of the flint. But this neat naming is to overlay a classification on the strangest of effects. And it is this. It's as if

a strike momentarily turns the hard flint to something like jelly, with the impact rippling down the flake and freezing there, creating a characteristic set of features that can be read as an attack by a human hand. The scarring is permanent; the blow recorded for millennia.

It's the energy of the blow that does this. The hard strike sends shockwaves through the mineral grains, which rearrange themselves as the energy ripples out from the strike point like a stone cast into a lake. It also makes it possible to identify a flat-topped, hollow-backed, bow-fronted, crisp-edged flake by touch alone.

The humanly worked flint (in the form of tools and waste flakes from tool making) that lies on the surface of the chalk Downland fields is almost exclusively white, as is much of the naturally broken flint beside it. The blue-black and grey flakes and split nodules have spent thousands to millions of years exposed to soil and the elements, and this contact acts upon the surfaces to give them a glorious, glossy white lustre, with the look and feel of fine meringue. Occasionally these white pieces are speckled by rich orange 'iron staining', which is just that: little spots of the mineral iron that has seeped from soil water into the flint's weaker points.

It's the soil water, too, that affects the change from coloured flint to white, dissolving quartz from the surface until the outer layer becomes pitted – almost bubbled – on a minuscule level, creating an optical effect that makes the surface appear white. The blogging biologist who posts as Arne Saknussemm says it's comparable to the head of foam on a pint of lager. The flint has not acquired a new skin, as I once thought. It is still flint, just as the foam on the lager is still lager. Or, to be more colour appropriate, the head on a pint of Guinness is still Guinness.

Sometimes you can see the difference on a single piece of flint, as weathering or the plough smashes off patinated

sections, leaving a stark contrast between white and black.

Some flint looks blue, as the white – not fully developed – covers the dark grey and black of the natural colour. Some fields specialise in blue flint, and the mottled, root-marked faces look like fragments of cheap Willow Pattern plates of indecipherable imagery.

Only the surface flint is so affected, and not all locations cause the same degree of colour change. Humanly worked flint found on some sandy soils, for example, might have no change at all, and look freshly knapped. Tools found in river gravels can acquire patina shades of burnt orange and caramel.

The tools and flakes excavated on archaeological sites look as pristine as the day they were discarded, held tight in the ground and unaffected by the churn of processes above. But there are myths and I have heard them first hand. The flints of the 11,000-year-old North Yorkshire archaeological site of Star Carr, for example, were a shiny treacle toffee when unearthed. By evening they had turned cloudy, and in a week white, as a film of fast-forwarded time crept over them. The story is a good one, but like many good stories, it's not true.

How many billion flakes have been struck in the service of humanity? Quick scrapers, slow-sculpted strike-a-lights, micro-flakes and mega tools. It's rare to walk a field on the chalk and not find a humanly worked piece. The number must be finite, but I couldn't begin to conceive of such a figure.

And still we are making more. Today's knappers must

be careful where they throw their waste lest it get mistaken for prehistoric scatter. I watch them in awe, demonstrating a single strike, or showing fidgeting children how to make a small tool, with precision born of dedicated practice. Waste flakes shower down (each one bearing the characteristic marks of human assault) until a perfect little arrowhead (the people's favourite) appears, charmed by the knapper's skill from the centre of a nodule.

I can't do it. Not 'I've tried and I can't', I just can't bring myself to break flint. Those nodules – from Kent, anyway – are 85 million years old. Most are lovely. I know where there are great piles at the edge of certain fields and around the churchyard in the spoil of the newly dug graves. The grave flint is fresh and would knap beautifully, but I don't want to be the one to split almost 100 million years of a thing.

Once I gave three beautiful flint balls as a present: big,

small and a perfect in-between. Being hollow and having their natural outer cortex, or skin, they were both lighter and rougher than the recipient was expecting. He didn't believe they were flint, so he took a hammer to one.

'It was hollow. This white stuff came out,' he told me, disappointed.

The white stuff was flint meal, a chalky powder packed with microfossils including spicules from sponges, tiny single-celled organisms called foraminifera, and even 'ghosts' – fossils that have so faded and degraded that it's no longer possible to know what they once were. I, too, was disappointed: sorry for the loss of a perfect ball, and sorry for the Kentish 'ghosts' swept away as dust on a foreign shore. I blamed myself as the agency of its destruction, and in no small part the recipient for his lack of faith.

Some of my best friends are knappers, and some of them have knapped me lovely things. Though there's a terrible tension for me between the natural and the knapped. Breaking a nodule is like killing a ghost fossil, and I find it distressing, although I have never confessed this to them. I often want to shout 'No! Don't break it!' before the first glassy-sounding strike. But the strike happens, and another piece of flint begins its journey from nodule to grain of sand.

Each of those pieces of flint, wearing their new flake form, will outlive me by centuries, and may perhaps enjoy the liberation. I tell myself this as I pass the shattered detritus on my friends' demonstration groundsheets, or

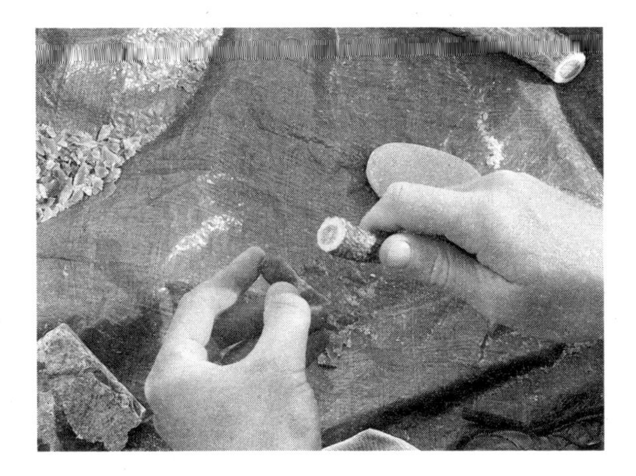

when I see the unwanted debitage swept into buckets for safe disposal.

Whither, flint?

Well, for now, as much of it in my house as possible.

chapter five

CHALKY LUGGITT

'Found anything?'

A couple in late middle age hailed me from the valley path and I picked my way across the ruts towards them. The woman held the lead of an angelic-looking Westie, who stared at me through the wire fence.

'What have you got, then?' The man was keen to see.

I opened my hand and showed them one broken fossil echinoid, a waste flake and a pretty red pebble.

'The flake's probably around five thousand years old,' I said. 'There's loads of humanly worked flint in this field. Mostly Neolithic, some later.'

'Nothing Roman then?' The man was cheerful in his disappointment. The Neolithic – a period that began some 4,000 years earlier than the Roman invasion – clearly wasn't his thing.

'No,' I said, trying to sound apologetic. 'No Roman. But then, I only find flint. Detectorists have found Saxon stuff up by the woods on Mill Hill.' I waved behind me, beyond the tree line.

'Ahhh,' said the man. 'We live over the way,' he nodded towards the estate about half a mile to the east. 'Do you know, they found the skeleton of a dead Roman sitting up against a tree when they were building our houses?'

'Really?' I said. 'A dead Roman?'

'Yes!' said the man. 'He had an axe in his head!'

'Crikey!' I exclaimed. 'An axe!'

'Yeah, poor bugger. What a way to go!' He sighed, as if the Roman had been in his class at school. He must have told the axe story so many times, the Roman had become a personal acquaintance.

'D'you know, I used to play in this field as a boy,' the man went on. 'It's called Chalky Luggitt. Yeah, Chalky Luggitt. On account of it being chalky, and because the strawberry-pickers would stop here to rest the horses on the way to Covent Garden. Near Ludgate.'

'How incredible!' I said. 'I didn't know that.'

I didn't. It was brilliant to know the field had a name. But I was still trying to digest the dead Roman. With axe. And Covent Garden. But near Ludgate? There's a mile and a half between them. It was all a bit wrong.

Over the next decade I would hear those same two stories many times from different people. Lore so local it was centred on just one field and one thirties housing estate.

The Roman story was a favourite.

'Yes, they dragged him up against the tree...'

'It was up the top of the estate. Found him in a garden

when they built the houses...'

'Died instantly. Killed with an axe...'

I've wondered about the story for years since. If a Roman had died sitting up against a tree with an axe in his head, the following things had to have happened:

1. The Roman was fighting an Iron Age Briton or one of his own compatriots.
2. He was either sitting up against the tree when he was fatally axed or was dragged there by companions.
3. After he died, nothing happened. At all. For around fifty to a hundred years. No one buried him. No animals came by to scatter his bones, no harsh weather, just the gentle disintegration of his body in place – legs outstretched, torso and skull concertinaed down, axe last, then tree felled or fallen without disturbing bones as they were gradually blanketed with soft, non-intrusive organic matter.

That would be a phenomenal sequence of events.

The day was cold and bright: almost warm in the winter sunshine. Having said goodbye to the couple – and thanked them for the local history – I carried on walking along the path that bordered the fence until I reached the west perimeter track. Clay had built so thickly on my boots that it was getting hard to walk and I paused on the grass by the path's edge and prised off great slabs of it with my trowel, checking for microliths in the peelings.

If it takes 10,000 hours to become an expert in something, I'm an expert in this field. And on this particular field. It's roughly trapezoidal in shape, wide edge at the bottom. Lying on the north side of the valley, it slopes up to a square of allotments next to a copse, through which it's a three-minute walk to the flint church and private graveyard of the Lubbock family. To the west lies the old, sunken turnpike road; to the east a more modern iteration that cuts just as deep into the chalk.

I have walked this field in every season and weather. I have chipped fossils from frozen clay till my fingers burned, sat drinking coffee from a flask among papery summer poppies in the years when no crop was sown. I've done circuits in cold drizzle, knowing that this might be the only chance I would get before the field was ploughed and seeded for the next crop.

I served my apprenticeship here, picking up and inspecting many, many thousands of pieces of flint, looking, feeling and judging them for platforms, bulbs, whiskers, scars. When

I began fieldwalking, I would take small bags to a London pub where a lecturer friend, over a pint, would perform a feat of miraculous prestidigitation on the tacky table, whizzing waste flakes away from tools and separating naturally broken shards, so that it became immediately, naturally, blindingly obvious what was humanly struck and what not.

Through his spell of seeing, I learnt to recognise retouch, hinge fractures and platforms of all periods. Here, in the pub, I picked Bronze Age from Neolithic, with some Palaeolithic thrown in. There was Iron Age, too, but their flint was shit. Large, crude and probably for rough work only. By the 1st century BC, the art of flint toolmaking was dying and I could trace its evolution and demise in that single field.

I'd just scraped off the last of the clay when I saw a man coming down the path towards me. He was wearing a light jacket over a shirt, as if he had just popped to the shop across

the road for a newspaper. We both paused to exclaim on the loveliness of the winter's day and fell into conversation. He drove and ran the mobile library that stopped every Saturday in the nearby village, he told me, and he was interested in owls. So when he locked up for lunch he would walk the perimeter path, looking for signs of the owls that were reputed to live in the nearby woods.

I showed him my finds and told him of some of the other flint tools I'd collected over the years. He was politely – possibly genuinely – interested.

I like owls but don't know much about them. I resolved to check for owl pellets under the trees next time I came. I didn't want to detain the man on his lunchbreak, so I said goodbye and walked on across the field to the first of five small rounds of trees, with more purpose than I actually had.

These rounds were planted as part of an 18th-century scheme of landscape improvement. Although tiny, each has developed its own character and is odd in its own way. The most westerly is a roost to parakeets and has a particularly handy smoking log – one I use often. The parakeets in the trees scold me furiously whenever I pause there for a cigarette, but they never budge from the branches. I'm told they're mesmerised by smoke.

The next round is thick with undergrowth and barely penetrable. Not to the deer, though, as I've often seen them at a distance, melting in and out of its bushes. Then there is the one that had the toolboxes. They were set below a large beech to one side of the copse. Who would carry toolboxes all that way to a tiny wood in the middle of a field? Did the farmer leave them? They were there just one summer, then they were gone. I never looked inside.

There is the one with the fallen trees – three, to be precise – showing a mass of topsoil and chalk in their roots, studded with broken flint nodules.

Within the most easterly – and the largest – one year, were a makeshift tent, an old plastic chair and a square church-hall-style table. It had the sense of habitation, and not that of a local kids' den. I didn't venture close. Some months later I peered into the copse again. Nothing had changed but it was clear the tent was empty. A year later only the table remained. I now think that I should have left a box of food, but I was afraid.

Having no cigarettes and no real business in any of the other rounds, I wandered into the copse of the fallen trees. I checked the tree roots for pirate treasure – as unlikely a thing I would ever find in the fifty miles from the closest sea. There was nothing, naturally, and I walked out of the round on the eastern side.

At the top of the field, near the long grass and brambles by the allotment fence, were two men; one had laid his detector down and was digging with a spade, the other was leaning on a bicycle. I walked towards them, and when they saw me they stopped their conversation and stared.

Hello! I said, with a small wave of my trowel, a kind of

declaration of allegiance, like a medieval pennant.

'Afternoon.' The detectorist was surly; he started digging again. The bicycle man was the talker.

'Archaeologist?' he asked, nodding at my trowel. 'Find much?'

I showed him the same three items that had failed to impress the friend of the dead Roman, earlier.

He looked pityingly at my fossil, pebble and flake. 'You want to get yourself a metal detector. You can pick one up quite cheap from Biggin Hill. Much more interesting,' he went on. 'All sorts of stuff around here. Frank here found a Saxon coin a month ago. Loads of them, up by the hill.'

Frank glowered. He had stopped digging and was picking over the earth.

'Do you know the hill?' Bicycle Man's question had a conspiratorial air. 'Live coven,' he mouthed. 'All sorts involved. Very active. Ever go up there?'

'No,' I said. 'Rarely. I'm interested in the Neolithic.' As if that precluded me from going up the hill.

A live coven, though. It was a day of revelations. I wasn't sure what I thought of this one, nor why I was being told.

The detectorist now had the interior quiver of the mightily annoyed, but whether with me, Bicycle Man, his equipment or his finds, I couldn't tell.

'Pahhh. The Neolithic,' continued Bicycle Man. 'There's a whole settlement here.' He pointed to an area behind the parakeet round, near where I'd been talking to the librarian.

'You screenshot an image from Google Earth, turn it into a jpeg, drag it into Photoshop and play with the colours, then you can see all the round houses.'

'Gosh!' I said. I was expecting long houses.

'Yeah, really whack 'em up. Bring out everything that's invisible on the image.'

'That's fascinating,' I said. 'I must check that out.'

It all sounded like bollocks to me (and proved to be so when I tried it). Also, I was not happy. There was something about these men. I made the usual calculation: my strength (limited) plus my weapons (a trowel and a very poor haul of flint) plus my visibility to passing traffic (should anyone be looking) and my ability to run to somewhere safe, versus their strength plus their weapons and speed plus their intentions...

I knew I was okay, but something wasn't right. There was a tension between them. I had interrupted something. So I pretended to be impressed for a few minutes longer then said goodbye, heading beyond them to the eastern perimeter path.

Most detectorists are charming. I have met several in this field and even exchanged details with a couple. One turned up at my door with a beautiful, deep-brown, shiny cow's jawbone, which he told me was the mandible of an aurochs – a huge, extinct wild ox. It wasn't, but it was a romantic story, and I didn't want to disabuse him. He said it had been found in the marshes beyond Dartford, which lent Dartford a dreamy feel. He also gave me a piece of flint quite unlike any I have ever seen. Polished – again – the colour of ginger cake, about six centimetres long and somewhat phallic with a tulip-shaped head on a thinner stem. It looked like something that would have come from an alchemist's cabinet, and who knows the life it had led before arriving at my door? I was grateful for the cow mandible and what I guessed was probably a fossil crinoid – a marine animal also known as a sea lily. Detectorists have no use for stone or bone, though this always amazes me. I offered the man a cup of tea and a biscuit and we talked for an hour about finds he'd made. Coins, mainly, as these were the most important to him.

As I crossed the field I wondered about the man and his coin collection and whether he was still detecting. I bent down to look at a flake, and glanced back. Bicycle Man was on his way out of the field, pushing his bike in the direction of the church gate; surly Frank was packing up his gear. Now it would just be me and the field again. Me and Chalky Luggitt.

I chose a plough rut and followed it. The clay surface was littered with flint of all kinds: small knobbly casts with beige-brown cortex; fossil sponge balls; hundreds of thousands of broken flakes in multiple shapes and shades. It's all pattern recognition, a constant register of similarity and difference. And over and over in my head the feeling I get when I see something interesting and translates into words as 'Oooh, that's a nice piece!' I have never been bored. Though I have on occasion been overwhelmed and had to sit down with it all swimming around me. Flint I have yet to look at. Flint that might be key to a whole settlement. Flint I shall never find.

One day I picked up an erratic – a rock that doesn't belong to the geological set of the landscape. Like a brick that is not Lego, found in the Lego box. It was lying on the surface of the field in a fallow summer when the land surface had become cracked and gravelly and grown chickweed and skinny poppies. Grey-green, pale-slatey, even. About as big as an iPhone but of course more curved, and with depth. One end tapered, the other flat. It was a tool. A greenstone – greenish-grey volcanic stone – pounder, possibly used to crush cobnuts. Or something. It did not come from anywhere round here and would have been traded from a distance.

I felt what I always felt. Thrilled and a bit embarrassed to be the one who collided with the thing after so many millennia.

All of history happened while that pounder lay in the field. Everything we know and were taught and all we don't know. The coming of metallurgy, the Roman occupation, the Norman conquest, cathedral building, Henry VIII's six marriages, the Renaissance, all the wars, Mozart, Picasso and the Moon landing. I wonder where the pounder was when Neil Armstrong said 'One small step...' Probably half a metre from where I found it and a hand's depth under the ground. Ploughing doesn't shift things much. These artefacts lead slow lives.

I will never know who made it, but I can name one of its historical owners: Bishop Odo of Bayeux. William the Conquerer's half-brother, and tapestry baron. He was granted these chalklands in 1066, along with vast swaths of the rest of England.

The land probably looked much like the country he came from, with woods and fields and grazing sheep. Did he even visit or just pick up the rent? Given the charges of misrule later levelled against him, and his warmongering, he probably never set eyes on it. Odo was an ambitious man. Perhaps you had to be in those days, but he took it to extremes, defrauding the church and secular owners of their lands (complaint was registered in the Domesday Book) and engendering a fierce rivalry with a fellow Norman, Lanfranc, the Archbishop of Canterbury.

It's believed he commissioned the Bayeux Tapestry – made here in Kent – to big up his role in the invasion, so the people knew what was what. His version of it, anyway.

'Scandalously immoral,' says *Encyclopedia Britannica* of Odo, who operated out of Dover Castle and is judged today to have been the richest non-monarch (with riches scaled) of the last millennium; it seems he just helped himself to as much as he could get. There was a rumour that he planned to make a bid for the papacy in Rome, or to seize the English throne from his half-brother – contemporary sources are anecdotal and unreliable. But ultimately Odo was imprisoned by the King for embezzlement. I pitied his poor tenants. Indeed, the whole of the country under the rule of William the Bastard and his men. I bet even the land hurt.

I crossed a line of large nodules jutting just above the surface. The chalk in which they sat was smeared across the thin soil on either side. The farmer had ploughed right down to the bedrock here.

A couple of metres to the south and downslope, the clay was thick and red again. And scattered with winkle and oyster shells plus the occasional iron nail. Remains of the lunches of the fruit pickers who worked these fields and the others down country, and who would have named the field Chalky Luggitt. Maybe the squared nails came off their carts. Or even, the smallest ones, off their boots.

I crossed to the most easterly corner. The clay was really sticky in the two metres or so closest to the valley fence, so I walked a little higher. Within two minutes I think I counted twenty winkles. I imagined the pickers there. Dirty, tired, laughing, bickering, singing. With wagons and horses and small children. Gathering the fruit and vegetables for a city and the hops for its beer. My great-grandmother and her family used to come from the Elephant and Castle to pick; grandparents on the opposite side had a pub in Hawkhurst in the centre of hop country. It had a sign above the door that read 'No Gypsies'. The pickers would take anything that wasn't nailed down, my mother once said. They even took a toilet seat. I had some sympathy for them and an indistinct memory of sitting in a field wearing stiff cotton, eating gritty bread. Maybe I read it in a book.

Small clouds appeared. I had been in the field for more than three hours and I needed to get home before dark. I looked to the east to where the sun had sunk behind a ribbon of cloud, turning it purple and frilling the edge with gold. Below the tree line on the darkening perimeter path, two walkers – indistinct in the shadows – were deep in conversation. One was tall and thin, the other looked like the librarian I had met earlier. They had a dog. Then all three stopped. One pointed skyward. I heard it before I saw it: small, from my viewpoint, but unmistakable. An owl. It rose suddenly above the trees then dipped and was gone. I stood watching for a minute afterwards, but the moment, the owl, and in the dark-rise the people, had disappeared.

I needed to go. The edges of the field were getting misty

and my fingers numbing slightly. The clay was sucking on my boots. I headed for the gate. Patinated flint still shone out: big broken nodules amid small chips.

I bent to look at something round, the size of a tennis ball, but stone. Not a spherical sponge but flint, it seemed, with no cortex. Greyish. I picked it up. Slick with clay and heavy, it was not a naturally formed thing but had been battered into a round. I wiped it on my jeans but couldn't get the mud off. I had some water in my bag and a bandana. I drizzled water on it and made an attempt at cleaning. It was a ball, not knapped but chipped, with indentations from continued strikes. Minute, almost toothed, mountains characteristic of hammering. It was worn a little smoother on one side as if rubbed or bashed on wood. In fact, that's what it looked like in part – the mashed end of a tent peg.

This was a hammerstone – someone's precious tool. Iron staining veined the grey. It was damp and smeared where I'd tried to dry it, and cold in my already cold hand, but it looked visceral. Sweaty, even. I held it to my nose. I touched it on my cheek. I wiped a smooth bit and held it to my tongue. It smelt earthy-irony. And behind that an invisible smell of sweat and toil and dog and shit and deep humanity.

Its shape made dating hard. It could have come from any time, though common sense suggested it belonged to the rest of the assemblage. At the least, it was 5,000 years old. I dried the remains of the wetness on my coat sleeve, then on the legs of my jeans, and looked at it again. In the fading light of the afternoon it felt melancholy. It was the weight – and colour – of a small human heart.

'It's all right,' I wanted to say across the millennia. 'I found it for you. I have it now and I'll keep it safe.'

I swung the bag off my shoulder and went to tuck it inside. But I couldn't slip it out of my hand.

Out of the field, crossing the perimeter path and valley road to the shadowed woodland track, I gripped the hammerstone tightly. I felt like a winner and a thief twice over, stealing not only the thing but the last traces of its ghost owner. But then, perhaps it was meant. I kicked and scuffed gobbets of clay off my boots as I hurried up the track looking only forward as either side of me soft things flitted in the waking dark. The hammerstone was warming to my hand. Chalky Luggitt had given me its best gift yet, and I never, ever, wanted to let it go.

chapter six

FIELD OF THE GHOST RIVER

Philomena was having the worst date ever.

I met her – and Justin – in the field opposite the farm, on the same sunny side of the valley as Chalky Luggitt.

Except today it wasn't sunny. It was warm and damp in a September way that made one sticky under the clothes and caused hair to frizz. Philomena's artfully straightened hair was frizzing painfully.

'Hi there,' called Justin loudly as I approached. He swapped his detector from right to left and shook my hand firmly.

'Great day for it! Archaeologist?' He nodded at my trowel.

He was tall, dark, handsome, in his late thirties, with a close-cropped beard. He had a lovely voice.

I told him about some of my flint finds. Plus the only metal of interest I'd ever picked up – a Saxon strap end that was so rusted as to be barely recognisable. I found it on a path's margin, just exposed from a low bank of soil. I reckon it was detectorist's discard – too ruined to be identified by anything but silhouette, too far gone for restoration.

Justin had not found anything yet, he told me, beyond the standard squared-off nails and a small horseshoe, which had come off a farm worker's boot.

I smiled. 'I used to think they belonged to tiny baby horses,' I said.

He laughed loudly. It wasn't funny, but I felt charming and witty for having said it. Which was clearly not what Philomena was feeling.

Justin was from Greenwich, and a former student of medieval history. Now he worked in the City. This wasn't his usual detecting spot. He was more of a Thames mudlarker, but they thought they'd have a drive out. He was field-testing a new detector. 'Philly' had his old one (still a pretty good model, by the looks) and was waving it damply over a flinty rut.

She looked miserable. She was quite beautiful and wore jeans awkwardly. Was she part of the field test? I imagined she'd pass the restaurant module, the party module and the weekend country-house module, but this was looking like a fail.

It was hard to watch and I didn't give them long, either in the field or out of it, so after a few more pleasantries I wished them luck with their finds and made my way to the upper track.

Can you have a favourite field? If Chalky Luggitt is my first love, this field of unknown name is my enduring passion. Neatly rectangular, it slopes from the hedge in the valley up to the track at the top, which, judging from the number of stone tools and flakes scattered across its beaten

surface, was a common routeway in Neolithic times, or even earlier. The track runs parallel to the valley and is bordered on the upslope by a thin strip of woodland. The other side of the trees is the track's twin, beyond which another field of similar proportions – one that is never ploughed – stretches up to the ferny heights of Mill Hill.

There is not a season, a month, possibly even a week – spread over the course of a decade – that I haven't walked this field. Or wanted to walk it. I think it's magic, but the magic is just nature where I've been lucky enough to tune in deep. Some years the field lies fallow, bearing scratty weeds that are, on close inspection, small botanical marvels. I've seen a New Year full moon rise in this field and been so awe-struck that I left too late and had to walk across a meadow and through a graveyard in the dark. I stripped off my clothes here in one warm October, when ladybirds fell in showers from the trees by the track and found their way into my underwear. I've seen ground frost rising from bare earth into little rills of steam that drifted east. When it's ploughed and seeded the earth looks like expensive corduroy. When there's a rainstorm it smells of 'dry', and iron, and mushroom and leaf mould. When it's dirt naked and the sun is low, the rays catch the gossamer of a thousand spiders woven across the clods, revealing a golden fairy path half a pace wide that stretches from my viewpoint into the blaze of the sky.

Its human history is long. If I were to follow the track west – which modern fencing has made impossible – I would reach first Caesar's Camp (something very, very old that Caesar had nothing to do with), a Roman mausoleum, then Keston, with its earthwork and its ponds. This leads on to Hayes Common and Wickham Court, both of which have traces of settlement from the Neolithic to Roman times. But all these lie on the fringes of the chalk and are as another land to me. Every year takes with it another of my broken promises to explore them.

My field has just one stand of trees, in the centre, and this, together with the land's undulations, makes winter days look like a Paul Nash painting. The shifting light sometimes shows the old river terraces – soft now – that step down to the valley floor where the road runs today. In low sun I find this mesmerising and try to picture the lost river as it flowed over time, high in the deep past, washing each successive floodplain with gravels and carving lower into the chalk as it grew old.

In my childhood, vestigial waters still flowed along the river's course as a nailbourne: a seasonal stream that flooded fields and houses. I remember driving through its eastern end with my dad in a Land Rover, creating slow, sloshing waves on the lane and going goodness knows where. Much of the water was run-off from the sloping land, and works have since funnelled it into large, purpose-built drains in the chalk. One such drain lies at the crossroads near the valley farm. The farm itself once drew its water from a well, now sunk eighteen metres below the ground. I believe the well water is that of the hidden river that still flows in a deep chalk cavern.

Were a slot to be cut across the valley and excavated, I would doubtless find older flint tools such as the toffee-shaded handaxe of *Homo heidelbergensis*, now under study by Frank Beresford. Assuming such finds, the oldest tools would be on the uppermost terrace, lost when the river was young, and the youngest at the bottom, where the older river flowed. Hominins from different eras hunted and fished on the banks of various incarnations of this river, and their lost flint tools are being used to date the river gravels when no other evidence is available: artefacts so old they have become geological markers. A terrace of pear-shaped early Stone Age axes might be 500,000 years old; triangular middle Stone Age axes 400,000 years old; more pointed Neanderthal tools 300,000 years old.

Each terrace, it's assumed, represents a span of around 100,000 years. When I walk the track that runs from the valley road to Mill Hill, I'm climbing terraces that take me several hundred thousand years back in time, if I counted only the geology below the land surface. As it is, all I actually see is the 85-million-year-old Cretaceous rock (in the form of flint) and rare traces of the 5,000-year-old Neolithic workings on the rock's fragments.

It was late summer, and the trees in the copse were thick with leaves. Years ago, I dreamt I found a polished stone axe a few paces to the north: smooth, slatey blue and perfect. I like this future hope, as my looking for the dream axe has found me all kinds of treasures there: a broken polished quartzite pebble with a hole in the middle known as a macehead, a tiny pink-tinged hammerstone and a beautifully knapped strike-a-light – a tool the size of a middle finger, knapped in the round with a blunt point at one end and sharper point at the other. The strike-a-light was a thrill. These tools are sometimes found beside patches of iron stain in Neolithic and Bronze Age graves, suggesting they were placed alongside iron pyrite nodules that have long since decomposed. The flint was used to strike the pyrite, making a spark that would ignite dry moss or mushroom tinder. It was effectively a Stone Age lighter.

Some twenty paces east I found the stone beads and the 'human' toe bone. The beads are the size of a flattened pea and grey-brown in colour. They are probably stone but could be amber. An archaeologist friend told me to 'go back and find the rest of the necklace'. Believe me, I've tried.

On this particular Sunday, I had free rein. The summer's crop had been harvested and cut to stubble – bleached-out

stalks standing to ankle-height and cruelly sharp. They were hard to walk on as they gave reluctantly and crunched like the crush of a hundred snails. I've spent whole winters with their scratch-scars on my wrists, as they score lines every time I reach down to pick up another flint, and another flint, and another...

There's no doubt they lived right here, the strike-a-lighters and the toolmakers before them. That their houses were here five millennia ago. It's a good spot. There was water, a sunny aspect, and abundant raw materials. What did they see as they looked around them? The real river, certainly. Deciduous trees, most probably, stretching up from the north-facing bank, from the farm paddock to the ridge. Sheep and goats likely grazed the clearings, which deer and pigs made easily as they browsed forest shoots. These people had dogs, and houses of wood, and hearths and pots, tailored clothes, and jewellery.

It's hard to conjure the vision and make it solid, like making an identikit face from just an ear. But I want it to be solid. I find the people's utter 'goneness' oddly hard. On misty days, when I can't see the flint more than three metres in front of me, I've half-expected them to walk out of the fog, or at least to see their shadows projected on the droplets like some juddering Super 8 film. Would they appear to me cut off at the knees, walking on their own land surface now buried under half a metre and a thousand years of accretion, or would they drift above the land in the patches long since scoured almost to the chalk?

However they manifested, they would not recognise the land that I walk. Our countryside looks ancient but it's a modern construct. I used to think of the bounty of flint available to prehistoric people on the Downs, but they would never have seen so much land with its skin off. Their hand-ploughed fields were tiny and if they wanted good flint they had to dig for it – which the people of my nameless field didn't do very well. The tools found here are mostly poor quality, solid in technique but let down by material. They didn't dig deep to the frost-free layers for their nodules, so there are all kinds of faults as the knapper's strike was derailed by flaws in the flint. I wonder why. The tools still worked. Perhaps they were bodgers like my dad. Perhaps 'good enough' was good enough.

I sat down and pulled a flask of coffee and a banana from my bag. Once, on a summer's day, I had snuck from the track through the tramlines where the crop hadn't seeded, and sat beside thigh-high barley to enjoy the exact same meal. On that day, as I sipped the coffee and picked at random pieces of flint with a fingernail, I saw something I'd only ever read about. A small, lozenge-shaped object that looked like it was made of sand. It was about as big as the top of my little finger, and it glittered in the light.

It was a crow pellet.

It wasn't sand, but tiny glossy snail shell fragments, pink and brown and yellow. I turned it over. There were two minuscule bones – mouse or shrew, probably. This was all the crow had eaten and couldn't digest – like the better-known owl pellets that are full of fluff and bones. It might have been regurgitated dead snail and crow spit but it was beautiful. It made my week and I have it in a box at home, wrapped in tissue paper and labelled in ink.

Break over, I carried on walking towards the round of trees then skirted the southern side. The earth was gravelly with beechnut husks that lay thick beneath a huge tree. A patch of crop into which I was walking had not been cut with standard efficiency but stood stiff and brown, the zombie stems around half a metre tall.

Something flickered by me, shoulder height. I stood very still. Another flicker. Over my head. I looked across to where Justin and Philomena had been. Gone now. No dog walkers or runners within my sight. Another flash in front of my face. I breathed as I realised that birds – small ones – were darting through the crop. They were moving so fast they were almost blurred to my sight.

Then they were all there – tens, possibly as many as a hundred birds, flying spirograph patterns into the trees and around the crop stems; diving for insects I could neither feel nor see. I was invisible or at least irrelevant to their purpose, as they skimmed my head and ears. I sat down on the ground. They flew lower, and so by their habit and half-caught shape I guessed they were swifts, bound soon for the southern Sahara. I don't know how long I sat there. Long enough, certainly, to lose all sense of everything but their woven flight and just perceptible 'screeeee'.

Then they were gone. They didn't fly off. They were just no longer there. I stood, stiffly. The sun had burnt off some of the sticky mist. I took a moment to refocus. I had been more than a witness; I felt like I'd been invited to a one-off event, a guest at a strange fairy dance.

It was a one-and-a-half-mile walk home and I started for the valley path, still a little dizzy with the after-effects of the birds. The sun was warm now on my back and the dampness was becoming steamy. I bent to pick up a couple of spherical, rough-cortexed sponges.

As I stood up I saw a dog running towards me. It was Colin the spaniel, followed at a saunter by his minders. Kevin was around ten, and if I ever knew his dad's name, I'd long forgotten it.

'Found anything interesting?' said Kevin's dad.

'It's all interesting,' I replied and smiled at Kevin. He was holding a large sponge and a nice echinoid. The sponge wouldn't fit in the pocket of his anorak. I could spare a little more time. Sometimes it's nice to be with people. I decided not to tell them about the birds. 'I've just seen some swifts over there,' seemed banal.

'I've got a bag if you want it?' I said to Kevin, and pulled an old orange Sainsbury's carrier out of my pocket.

'Cool, thanks!' Kevin, took it and put his finds in the bottom.

I showed them the handful of flint flakes I'd found – the flakes were only debitage from knapping, none of them tools.

'Have you done the top today?' Kevin's dad gestured to the eastern corner. Colin the spaniel ran around dizzyingly. I knew they didn't like him too near the road, so we walked up to the track then doubled back, walking in a line a pace apart, taking slow steps while precision looking.

I'd met Colin, Kevin and Kevin's dad a couple of years ago, walking that same field, just as a thunderstorm was about to hit. I was walking a few metres in, and they were coming up the track from the valley. The sky was black to the west and, worse, blacker overhead. Getting wet was not a problem but getting struck by lightning was. We all three ran to the shelter of a nearby wood – it was a risk, but less risky than being the highest thing in the middle of a field, and while the rain poured and lightning played, I showed them my finds of that afternoon.

When the storm ended and the sun came out, we went back into the field and Kevin proved pretty good at picking out striking platforms and bulbs of percussion. He went home with a dozen waste flakes and a scraper, plotted on a spare – and now damp – photocopy of my finds map.

We met occasionally, their dog walks colliding with my field walks, and spent a bit of time in companionable looking.

Today, we were doing well. Kevin found a partial echinoid embedded in a nodule, a piece of pot that looked fairly modern, and some mineralised bone with nothing to suggest what animal it came from, though I reckoned sheep.

I found some more waste flakes, a small broken tool with nice, serrated retouch down one side, and a piece of flint with some pretty crystal in it.

Kevin's dad found a large pebble. I checked it. There were no signs it had been used for smoothing or hammering but he was pleased with it.

I reckoned I had another half hour before I needed to start for home. I usually leave it too long. Just one more rut, I always think, and then the sun goes and the dark rises. We'd come halfway down the field and it was probably enough. Kevin's dad was way ahead with his pebble.

I was loosening a fossil with my trowel for Kevin and we both glanced up. Kevin's dad was about five metres away, standing. And staring into the valley.

'Look...' he called back to us quietly.

We stood, abandoning our stuff, and walked towards

him. The road was not the road any longer. While I could still hear the swish-clunk of the eastbound cars as they hit a loose drain cover, I couldn't see them through the hedge gaps. I couldn't see the hedges. Mist had risen from the valley bottom and was lying softly, dense, even, and gently roiling about four metres from the ground. It had become the river. As we watched, tracing the mist's snaking path east and west, the low, hazy sun sank closer to the cleft, turning the pearly mist rose gold. The ghost river had risen again, and we watched until it flowed neon pink into the setting sun.

This was the river our flint knappers knew. We stood silent – even Colin the spaniel – as the mist lapped at the fringes of the fields above the hedges. Altering the landscape in ways I couldn't ever have imagined. Drowning the paddock, marooning the farmhouse on an island and spilling over the faint lower terrace I had walked a hundred times. It was as solid an illusion as I have ever seen. And as it rose, I choose to believe the people of the field rose too, and stood behind us, near to their river's edge. As long as I never turned, they would stay behind me, close to me, just 5,000 years away, and I knew while I could walk this field and find their lost and forgotten things, they would always be there.

On the far bank, in a clearing but indistinct in the fading light, a tall, rangy man was walking, head bowed and hands clasped behind him. He paused and looked across at us and raised his hand in salutation. His dog paused, too, then trotted on.

chapter seven

STRANGE MAGIC

It was January, and I was coming down the Field of the Ghost River, walking slantwise from the small stand of trees to the valley floor. The field had been turned; short lengths of last year's crop lay crisp and dead on the surface. The soil was speckled with chalk, and the previous week's rain had washed the top flint clean.

I'd crossed a good three quarters of the field when I came upon a Thing of Difference. A sooty flint nodule lay black against the grey/white of the soil. I bent down to touch it. The cortex wasn't the browny-black manganese you sometimes get, but a deep, powdery black. It was burnt. I picked it up. Black powder came off on my fingers. About a metre away, I could see another piece. This wasn't just burnt but melted. The shard was jagged and sharp, suggesting it had shattered violently before turning glassy and bubbling on one side. To the left of me were some charred stalks. I walked up and down for a bit. Whatever had happened here had happened before the very recent turning of the earth, as the stalks and fractured flint seemed mostly to be folded back in. I scraped at some earth beside the stalks, and then some by another piece of flint. It was carbon black.

I walked back up the field to look down on the area and could see a few more burnt shards. They covered a patch that was sort of oblong, motorbike sized, if the machine was laid out on the ground. I don't know why that analogy came to mind.

I already had a piece of fire-cracked flint with me that I'd picked up at the field's western edge, and know burnt flint from garden bonfires well. The piece I had was typical: blue-grey and crazed with tiny, thread-like veins. Sometimes it's pinkish, sometimes grey-white. Sometimes, if it has been in the fire a long time, its texture goes almost porridgy. Sometimes it looks like blueish china. If it's worked flint, such as a core, it will twist and shift from its original shape just enough to make you doubt what you have in your hand.

This was not that. This was not bonfire flint. This flint had been subject to temperatures hotter than hell.

I looked at the patch again, and the wider context of the field. While I will never be sure, I firmly believe I had just walked through the aftermath of a lightning strike.

February, and the weather had turned from [illegible] to wintry. Snow covered the fields and the paths were icy. With walking curtailed, I set off with my niece on a trip to Tunbridge Wells – more specifically the Tunbridge Wells Museum and Art Gallery. The train journey from Chelsfield took us through deep cuts in the chalk and three long tunnels. At Knockholt Station, great horizontal bands of flint were visible in the vertical railway cutting, but after the Polhill tunnel – that runs for nearly two miles beneath a high ridge of the North Downs – we were off the chalk and in a land underlain by mudstone, sandstone and clay.

The museum is on the first floor of Civic Centre, a handsome 1930s neo-Georgian building. The room we were looking for was unlit, to protect the taxidermy display from damaging light. The two huge, west-facing windows were masked with putty-coloured blinds that allowed just enough light in for us to make out the exhibits, and no more.

I needed more. In the case between the windows was a modest collection of archaeological artefacts, and the treasure I had come to see. Except I could barely see it. I experimented with the blind on the left. It was fixed and I couldn't raise it, but I could, without creasing the fabric, pull it aside a little. A pocket of colder air puffed into the room. Outside, it was snowing: big white flakes from a yellow sky.

I found a pen in my bag and used it, horizontally, to prop open the blind a little, the blind's weight holding the pen fast against the glass. It would have to do.

From the hand-typed label I read:

> *Finds from an Iron Age burial*
> *This pot containing a fossil sea urchin and a Neolithic axe-head were found in 1910 as part of an Iron Age grave in Powdermill Lane, Tunbridge Wells*

Beside the label, on a small Perspex shelf, was a small, brown, urn-shaped pot, the fossil – a white, heart-shaped sea urchin, a micraster – and half a partially polished flint axe. The axe was about as big as the space made by my forefinger and thumb if I held them up into a U.

I took dull photos from every angle. I made small nose-smudges on the glass of the case. I stared. And stared. My niece wandered off.

A chapter of a book I'd read previously called *Myth and Geology* had told me what was on the museum's catalogue card in the archive, and it was not much more. The pot was found in August 1910 by two workmen named Mercer and Crittenden, two-and-a-half feet below the ground, while they were digging the foundations of Hillgarth (a house that was later demolished to create a small close of houses) at the corner of Powdermill Lane and London Road. Three feet away was a large, covered urn, at least fifteen inches wide, which contained burnt bones. The urn was retained by the house owner and was later broken and lost. The small pot was dated to around 300 BC.

I couldn't suck up any more information by staring at the museum case and had questions that would never be answered; not least, what happened to the cremated bone. I retrieved my pen from the blind and my niece from the terrazzo-floored foyer, and we left the Tunbridge Wells Museum and Gallery, walking out into the settling snow.

I'm just one in a very long line of people who have been bewitched by flint – a line that probably extends back a good 500,000 years. Some of the most magical pieces are found in the form of sea urchin fossils, which are perfect casts of their plated internal skeletons. Each fossil was formed when the dead creature's body was covered by chalky seabed sediment, leaving a hollow in the shape of the urchin as its body decayed and the sediment hardened around it. This hollow was filled with silica-rich water, the silica, over time, hardening into flint.

Urchin fossils charmed our Palaeolithic ancestors, and we have the evidence: a famous flint axe from the river gravels of Swanscombe in Kent was knapped around the cast of a *conulous* (cone-shaped) sea urchin. The axe is a beautiful burnt-caramel colour, the urchin's base sitting, like some magisterial *heidelbergensis* coat-of-arms, in the centre.

This hominin's tool was made around 400,000 years ago; there are others, too. An axe from West Tofts in Norfolk featuring a fossil bivalve (like the Shell logo) on a piece of deliberately retained cortex. This one may be 500,000 years old. In France, hominins turned a heart-shaped fossil micraster itself into a scraper by judicious chipping of its lower edge. Much later, Neanderthals living by the river that we now call the Charente, in France, made a scraper from flint with an urchin fossil slightly offset from the edge, proving that fossil enchantment is no modern, nor modern human thing.

In the late Stone Age, fossil urchins – like the one I had just seen in the museum – were placed in graves, usually singly but sometimes in numbers. One early Bronze Age grave in particular is extraordinary for its urchin grave goods. It was discovered in 1887 in southern England, on the chalk Downs above Dunstable in Bedfordshire, when a farmer was levelling mounds in his fields. Worthington George Smith, an illustrator, expert in fungi and Palaeolithic archaeologist, excavated the mounds, and found the bones of a woman (he named her Maud) and of a child. But more extraordinary, the mound in which Maud was buried contained a hundred and three (or 147, or 300, according to who was doing the counting) fossil urchins.

Smith's drawing of the burial presents an evocative picture: the urchins, looking like hundreds of little cucumber slices, making up an oval frame around the folded bones of mother and child. The fossils' actual positioning was more random and possibly not exclusive to Maud's grave, though the image has somewhat refreshed the reality of the time.

We can't know what the urchins represented. What if she – and those in her neighbouring graves – just liked collecting fossils? I must have more than 200 on shelves and in boxes that would make excellent grave goods, should I choose to be buried. (What, otherwise, will my family do with them all? If I'm cremated, I've told them to chuck my collection back on the fields they came from; no museum would want them.) But for Maud and her companions, this explanation seems too trite. The fossils meant something, probably otherworldly and possibly of actual or talismanic value. They come up too often in archaeological contexts not to have purpose. It's just that the purpose can't be married across time with the circumstances of their finding.

Fossil urchins have been thought lucky through history. Known in English as shepherds' crowns or fairy loaves, they were pocketed by generations of finders in past centuries for protection, both in the home and out. They helped dairy workers churn milk into cream, prevented infants from being stolen by fairies, and protected against the demon horse, 'the nightmare'; and most especially, in northern lands, they repelled trolls. They were set upon windowsills, above

doorways and on mantelshelves, sometimes even varnished or polished with stove black.

Maybe Maud's fossils were lucky – for someone, if not for her. Or urchins could have been part of a wider system of belief and worship that stretched deep into the past, was known across Europe and Asia, and was practised right into the 19th century.

Because what is known, through rituals considered old even in Ancient Greece, written accounts, illustrations and some dogged anthropological study across Europe and beyond, is that for much of recorded history, fossils – and also stone tools – were 'thunderbolts' or 'thunderstones' and were gathered and secreted about the home as protection against lightning.

The theory went like this: flint axes, polished stone axes, maceheads with circular perforations, fossil urchins and arrowheads were the cause *and* the product of lightning. That stood to reason, as they were found in fields where lightning struck. Often bright and shiny after a rainstorm, these thunderstones had not been 'there' before and so it was clear they had fallen from the sky.

In the (erroneous) belief that lightning never strikes the same place twice, and in an action that was almost homeopathic in its logic, these fossils and tools were curated and stashed in the rafters of houses, above doorways, in corner foundations or on bedsteads.

Certainly, their folkloric journey was magical. They fell

to the ground with the lightning strike, plunging seven fathoms deep into the earth. Then each year they rose a fathom, until after seven years they surfaced, where the fortunate might happen upon them. There was nought to be gained by digging for thunderstones. It just made them retreat deeper into the earth.

There was nothing, it seemed, that a thunderstone couldn't fix or cure. Pieces could be ground into a powder and taken as a cure for rheumatism; they frightened rats from flour and corn and, when hung in the brewing vat, prevented spoilage by trolls. They were even, in some lands, worshipped as gods, and sacrifices of precious milk, butter and animals were made to the holy stones.

That fossils should be deemed magical is one thing: they are strange and beautiful and some look like the petrified animals they are – but not quite, and not all. They seem almost embryonic, partially formed; and while some Ancient Greek natural philosophers reasoned, rightly, that they were the remains of early sea creatures, others looked to a whole other philosophy. This philosophy went by the name of plastic virtue (using the original meaning of the word 'plastic' as 'malleable'), in which there is a great animating force in the world that moves everything toward life's creation. This vital force stalled in some instances and resulted in the partially lithified almost-beings.

But in the case of tools, it's harder to understand. Why did the mundane become magic? Why did the simple, repeatable technology of knapping and polishing enter the realm of the supernatural?

All 'whys' are speculative because the 'when' is so long ago. The thunderstone belief was so widespread that it's thought to date to very ancient times, gaining currency as people or culture, or both, took root in different parts of the world. It was present in the Eastern Mediterranean (where metallurgy was spreading rapidly) as early as the 9th century BC. And in Greece, Rome and Egypt, the protective power of amulets made from Stone Age tools was written about by scribes who seemed not to know the objects were human made, or not to care, if they did know.

Thunder and lightning mattered, too, in Iron Age Tuscany, where the Etruscans performed divination with their Brontoscopic Calendar (Bronte being the Greek goddess of thunder), with detailed predictions such as:

> *JANUARY 7:*
> *If it thunders, there will be a slave revolt and recurring illness.*
> *JUNE 9:*
> *If in any way it should thunder, there will be a loss of flocks through being overrun by wolves.*

Flint arrowheads set in gold have been found in their tombs.

So it is possible that, as people started making their tools from metal rather than stones, they quickly forgot the old ways of knapping and polishing and began to see the shaped stones as marvels. And after all, some stones, which we call 'aeroliths', do fall from the sky. They are jettisoned from thunderous volcanoes and fall as meteorites. The chances of an eyewitness among Earth's tiny ancient population is doubtful. However, most people would certainly have witnessed flint as a fire-starter, creating lightning-bright sparks with iron pyrites.

The whole cult is gone now, dying just as people had the means – and early folklorists the wit – to record it. Science and the Industrial Revolution saw off the people's magic. But I wonder if there are any undiscovered thunderstones left, tucked behind the beams in the roofspace of old houses or churches, or buried under the floors, protecting against lightning, trolls or more. I hope so. I choose to believe that all of them are lucky – the tools, the waste flakes, the pretty stones and the fossils.

But no matter how many urchins and knapped fragments I stash in the corners of my house and close to the rafters of my roof, it hasn't cured my fear of thunder and lightning, which grows fiercer with every year. As a brontophobe and astraphobe, I know it all. How the pine trees shatter and spear you if you're sheltering under one that's struck; how lightning's branches travelling earthward seek an upward streamer streaking skyward from a house, a bush, or your head in order to land the strike; how to get into the lightning strike position when there's no choice left (you perform a kind of downward dog, tucking your head under so the lightning hits your back and travels down your arms, missing your brain; you might then stand a chance). It is not good to be curled in a hawthorn hedge on an old field boundary in the rain, but it's better than running across the middle of a raw and wet clay field knowing that, if Thor strikes right, you will be ripped asunder and die where you fall. And if you survive? You will never be the same.

So sometimes I wish I had dug a little deeper in that blackened patch of field and found my own thunderstone for protection. Although, if the lore is true, the stone was seven fathoms deep when I crossed the strike and is only now beginning its journey to the surface, for some lucky passer-by to find in years to come.

It was a warm Saturday in April, and I was taking a long path, walking out of myself across newly turned fields that I didn't know well. The flint was becoming dizzying, and this particular field was more flint than clay, a good part of it stained and streaked with iron pan. As a child, I used to think the marks were fairy writings. Messages I couldn't read, not meant for me.

I was thinking about elf-bolts, or arrowheads, and how I had never found one in all my years of walking, though I had found just about everything else. They were small, admittedly; expertly knapped from quartz and flint. While thunderstones were not believed to be of this earth, elf-bolts or elf-shot had supernatural origins closer to home. In later Saxon times, elf-shot was a known medical condition suffered by people and livestock, which was caused, obviously, by an elf with a quiver of arrows, made by the ancients – the 'others' – who had somehow crossed the divide from human ancestor to fairy foe. Elf-shot caused all sorts of problems, as detailed in the Old English medical text *Wið Færstice*. The term *Wið Færstice* means 'against a sudden/violent stabbing pain', which, in truth, could have been anything, but was most commonly thought to be rheumatism.

In a dose of sympathetic magic, any elf-shot found and steeped in a pail of water might cure your ailing cow – or your ailing self – and amulets made of the elven tools might similarly protect you from fairy wrath. These amulets were often fancy, the tiny flints set in wrought silver to be worn as a charm around your neck.

I'd found nothing in the field I was walking but a couple of small fossil sponges. Now here was a third, small tulip-shaped fossil with a smoothish, sandy-coloured cortex. I bent down and picked the mud from each end with my fingernail. It was starting to look like a large, elongated stone bead. If there was a hole and it went right through, in folklore, this would be a hagstone, and a lovely one at that. One I could wear, and almost as delightful as an arrowhead.

I had a wooden chopstick in my bag as part of my toolkit. I picked the mud out of one end of the fossil, then the other. The possible hole was about a centimetre in diameter; the entire stone two-and-a-half centimetres round. It felt solid in the middle and the chopstick wasn't up to it. I got out a Swiss Army knife and eased a fine blade into the mud core as far as it would go. I poured some water into the space. And waited. I withdrew the blade and used it to pick at the mud. About ten minutes of drizzling and picking, and I had enough of the mud out to see that the hole went all the way through.

A hagstone is essentially a flint with a natural hole in it. Perhaps the flint formed around something that later disintegrated; perhaps it was weathered. Perhaps it was the fossil of a Cretaceous sponge, like the one I had just found. Most importantly, hagstones were considered lucky, and like thunderstones there seemed no end to their powers.

They were worn around the neck by men, passed down through the family line; they were spat on and wished on; tucked in a pocket as a charm against the spells of witches and eye of gypsies; tied on strings to cottage doors for the same, or on the bedstead to ward off illness.

Keeping one in the house meant the householders would never be in want; above the staircase meant safety from mishaps on the stairs. They were set on windowsills; used as key fobs; hung in fishing boats, and in dairies for the health and protection of cattle; and, in the early days of the motor car, lashed to the axles for luck.

I kept turning my find in my hand as I stepped out of the field into the narrow, tarmacked road, which was warm now and soft-smelling, and I passed the end of Bogey Lane – a natural, flint-cobbled track that rises between two fields. I've walked it many times, a cart-width lane that runs for around half a mile from farm to farm. Once, all roads over the Downs would have looked like this. The name always makes me laugh at the same time as it makes me slightly wary. Bogey, Bog, Boggart, boo...

But today I had my hagstone, so I was fine.

I followed the road a few hundred metres to the bridleway up into the woods. The April sun was casting spectacular tree-branch shadows onto the road, and I stopped to take a photo. Part of the shadows looked like a hare, sitting alert with long ears listening. I stepped farther back for a long shot, and the shadows resolved into a bigger picture.

Gone was the hare, and in his place was a great horned god riding some kind of beast. I looked above me, fully expecting Herne and his full Wild Hunt to come bursting out of the trees... But it was just twigs and branches shadow-painting the road.

I turned onto the bridleway and followed the path to the dene hole. The day was getting warmer and I could hear spring coming to the woods in a million tiny rents and tears as blossom and leaves burst their buds. The whole forest was crackling.

The leaves of the not-yet-bluebells looked squeaky green as I took the path to the dene hole fence. But something was there that was not there before.

In front of the railings was a circle of large, almost head-sized flint nodules. The circle itself was about two metres across. I felt strange about it. A bit indignant really. Who had made the circle and why? To me, these woods couldn't be more magical. Did they need a magic circle? I assumed magical intention – possibly magickal intention – for after all, what else could it be?

I stepped into the circle. I didn't appear to have entered another realm. The same birds sang in the same trees. I stepped out of it. Still in the woods.

I walked around it. Widdershins, of course. Then, out of badness, I looked around for elves. No elves. I looked through the hagstone, like a tiny telescope. I got some mud in my eye. I caught a glimpse of something. A flicker in the

trees. I lowered the hagstone. A squirrel leapt from a low branch in a liquid bound.

Maybe it was somewhere for someone to sit? A meditation spot? Though there was no trodden – or sat-on – centre. I looked at one of the nodules closely. I couldn't be sure where they'd come from, though the white of the cortex suggested they were not long out of the ground. I photographed them. There was nothing more to be done.

I took the path that led home through the woods. Everything was light and bird filled, pre-leaf. I turned my hagstone in my fingers. It felt nice.

People were about. Couples walking. Two boys on bikes. Some friendly dogs. There's a glorious tension in the April woods that almost hums. I was on the last downhill stretch when, ahead of me, a crow stepped onto the path. I threw a couple of monkey nuts from my pocket but the crow flew off.

I turned back as I reached the field gate. No sign of the crow.

It was still early, so I thought I'd skirt the field's edge, climbing to the top lynchet in the sunshine. The lynchet is a steep bank between two fields – the result of ploughing over hundreds of years as earth was pushed by the plough and banked up along the edge of the flat, hilltop field, and was in turn cut into by the ploughing of the sloping field below. The bank was about three metres wide and covered with a line of hawthorn bushes. There were a couple of scramble paths forged by animals that joined the two fields. The path half-way along was the easiest, though I would still need a runup and some quick bush stem grabs to get from the lower field to the upper without sliding back.

I was about to dive at it when I realised I couldn't. There was a cairn in the middle of the path. A flint cairn, and in the centre a great phallic nodule almost half a metre long, pointing skyward.

Someone had gathered the flint and arranged it very nicely, into a kind of coronet halfway up the bank. Was it a memorial for a dog? Or maybe a person? I was a little wary of what might be underneath.

It was rather lovely but didn't make any sense. Not that there was any reason why it had to, for anyone beyond the maker. Why had I never thought to make a cairn? Apart from liking the flint either with me, or where it was. I felt oddly proprietorial; but the flint wasn't mine. I was surprised to find that I somehow thought it was.

Later, at home, with my hagstone washed and laid out beside the kitchen sink, I couldn't shake the feeling that there was a rival for my flint affections stalking the woods. A rival who was making, if not magic, at least art. I found my copy of *Robinson Crusoe* and turned to 'The Footprint' to see how Crusoe felt when he discovered another person on the island:

> *'It happened, one day about noon going towards my boat, I was exceedingly surprised, with the print of a man's naked foot on the shore, which was very plain to be seen in the sand. I stood like one thunder-struck...'*

I read no further and closed the book.

I don't know how old I was when my father started keeping stuff under the house. He was always acquiring things and needed somewhere to put his finds. I'm not sure how I thought he'd got under there, but it turned out he'd dug a pit right through the concrete floor of the adjoining garage, next to the house wall, till he hit foundation depth. Then he took out the bricks that formed the outer foundation wall to make a 'door' into the crawl space. It was no cellar. The house was 120 years old and the space just two-and-a-half feet high.

He'd rescue lawnmowers from skips and love them back to life. This is what he first started keeping. I think at one time there were five Mountfield mowers – some of them classics – in various states of repair. For years we thought this was all there was. But it wasn't. Like some metaliverous Steampunk crocodile, he cached treasures from his garage and beyond: old half-empty paint cans, bicycle parts, two metal filing drawers and many Land Rover parts.

The Land Rover parts included two wings, stashed at the crawl space's distant edge. This is what I was there to retrieve.

The garage electrics – including the sticky web of wires and bulbs strung under the house – had been made safe by my brother, and they worked. Still, I took a battery-powered floodlight. My brother was in the garage and my mother in the house above, and while I could have gone down on a day when no one was around, it would have been foolish. What if, for example, I hit my head and knocked myself out? Or was eaten by a monster? I stepped down into the chopped-out entrance, crouched into a ball, taking care to tuck my head tightly under, and kind of rolled in.

Before me was a gritty plywood crawl board that ran up the central 'aisle'. It was placed on the bare scraped chalk, over which lay a dry brick, mortar and dust slurry. On either side of me were brick pillars, sloppily mortared and only a single skin thick. They created a series of 'rooms' that were filled with white plastic buckets – the kind that had held some sort of useful chemical, and that my dad would bring home empty from work. Some I could see into. Inside were tools, mostly random and unsorted. And metal spares. One small one was chocked with aluminium bolts. I recognised all the shapes of the spares even if I couldn't name them. Heaven – for some such as my dad – must be an endless breaker's yard, with every possible part for every vehicle made in the 20th century. I hoped he was there now.

The wings were at the far side of the house by the old back-room chimney. Johnny Cash was singing from a CD in the garage, a 21st-century sound system and remaster providing a mournful clarity missing from my childhood of mono LPs and third-generation cassette tapes.

I crawled past two headlights and an intake manifold. My dad must have spent weeks under there. Dragging this stuff down and arranging it. Pulling it out and taking it down again. It was the last space on earth that was truly his; we'd long begun the big sort of his eight sheds. The ninth had collapsed, the roof ruined by rain, foundations beset by badgers.

I was nearly at the far wall. The sound of vacuuming in the kitchen above me – my mother, cleaning – was drowning out Johnny Cash.

In my experience of Neolithic passage graves, knees give

out before courage does. I grabbed an accessible end of the first wing. I pulled it.

It was so light, I'd used too much force and the front corner hit me in the cheek. It hurt. The vacuuming reached a crescendo.

I hung on to the front of the wing and crawled backwards, manœuvring it along past buckets, boxes and the metal cabinets, until I was out of the entrance. I unfurled stiffly and wriggled the wing up and out. It took a bit of doing.

Now an expert, I stepped down, curled up and began the second crawl. Only this time, as I grabbed the front end, the wing didn't come towards me easily. The front swung round; the back was snagged. There was a smallish 'ear' where the wing bolted to the Land Rover's body and it was caught. I wiggled and shoved the wing but it wouldn't give. It was as if something was hanging on to it. I thought of the wiry ghost of my father gripping it in defiance and felt disconcerted. I gave it another two-handed tug and the thing relinquished its grip. The wing bounced up and hit the rafters, throwing up a massive cloud of chalk and house dust that rose and fell back in the floodlight's ethereal beam. I, too, had risen up with the ricochet and hit the back of my head on a very tangible beam. It hurt in a splitting sort of way. I curled forwards on my knees. My heart was beating in my throat as glittering motes rained down. It really hurt.

I was underwater, on the bed of a shallow Cretaceous sea, watching a scene from the eon-long snow of tiny dead things that would settle to become chalk and flint. But I couldn't stay there. Whatever had snagged the wing was now dislodged and coming into focus. It had been beside the foundation bricks of the chimney. I had the floodlight on it and, in the settling dust, could see exactly what it was.

A man's lace-up boot, minus laces – old, creased and chalky grey – stuck out of the dry floor, disturbed for the first time in a century. Papery-looking, almost corpse-like, it was a foundation deposit, for sure. A 'spiritual midden'. Not for protection against lightning, but against other stuff, such as the devil. Boots and shoes were spirit traps, attracting demons, ghosts and witches with their smell. Once in the boot, the spirits couldn't get out, and were imprisoned for all eternity. There was a saint said to have trapped the devil in a boot, but this was probably the Christianisation of something much older. Household deities from Europe to Russia also liked footwear and things like old corsets, hats, gloves and even visards – velvet-covered pressed paper masks, worn by Elizabethan ladies to protect their complexions.

And who could ever mock such beliefs? We're all hanging on to the physical world for dear life, doing deals with objects to better our own narrative and fortunes, and make wishes for ourselves, our families and our friends come true.

The boot had served us well and I had a duty to it. I crawled out backwards, in some pain, and went home, where I assembled: a beautiful patinated flint knife (originally from the field known as Chalky Luggitt); a fossil echinoid (of the fairy loaf kind); an old silk camisole; a two-pound coin and a headache pill. I swallowed the pill and returned to the garage. I fed the first Land Rover wing back down, then went under, pushing it before me until it nested with the second wing, then I pushed them both back to the wall, tilting the ends up to place the flints and the gold coin – wrapped in the camisole – next to the boot. I thanked the boot and saw to it that it was comfortably reinstated.

Footsteps above my head sounded like a giant walking in seven-league boots, but it was only my mother putting on the kettle. I began the backward crawl. Johnny Cash had paused. Then from the garage he began again:

> *'And I heard, as it were, the noise of thunder, One of the four beasts saying,*
> *"Come and see."'*

I hit my head on the house wall again coming out.

chapter eight

PARAMOUDRA DREAMING

When granite rose from out the trackless sea,
And slate, for boys to scrawl – when boys should be –
But earth, as yet, lay desolate and bare;
Man was not then, but Paramoudras were

From the poem *Specimen of a Geological Lecture by Professor Buckland* by Philip Shuttleworth, *c.*1822

There's a point, part way along Ramsgate's long, eastern harbour arm, that offers a perfect view of the town, and of the geology that gave the original coastal settlement its name. Or more truthfully, the lack of geology, for Remmesgate, as recorded in 1275, comes from the late Anglo-Saxon 'Hremmes' (from earlier Hræfnes) for 'raven', and 'geat' as in gap, or gate. It was as romantic a name as anywhere I knew in Kent, and referenced the shallow valley that cleaves the chalk cliffs rising on either side, some thirty metres above the sea.

The moon was silvery, setting, and on the wane. I would like to say it was coming down bang into the cleft of 'raven's gate' but that would not be true. It was hovering somewhere over the roofs of the Georgian terraces on the East Cliff, yellow in the lightening dawn.

I walked back along the harbour arm toward the moon and the town, slipped down by the Victorian pavilion – restored now and the largest Wetherspoon in Britain – and on to the beach. I had the sands all to myself. In fact, I had Ramsgate to myself. It was 6.35 in the morning on Christmas Day and I had snuck out of my friend's house like a thief to steal this precious time and feel like the only one alive. Except for the dog walkers. There were a few shadowy pairings on the esplanade. There are always dog walkers and I tell myself they don't count. And I was there for a different reason: to find something I could only see at low tide. That morning, I had a plan.

It was hard walking on the soft, dry sand, and I made my way to the solid strip by the ebbing waves. The wet sand was streaked damson in the dawn light, pooling the orange on the horizon and the multi-blue hues of cloud and twilight sky.

I stood with my back to the sea and looked up to the town. I was roughly in the place that the painter William Powell Frith would have set his easel to make sketches for his famous work of 1854. Two years in the making, *Ramsgate Sands (Life at the Seaside)* showed the sunny summer sands packed with Victorian society – of all classes – at leisure. The short, right-cast shadows suggested it was painted in high summer, around four o'clock, or perhaps later.

The painting – in photography terms – is a mid-shot. Dapper gentlemen in stovepipe hats and ladies in bonnets and vast crinoline skirts with pretty children stand right at the lap of the waves, their shoes and petticoats almost certainly wet. Bathing tents are pictured to the left, though no one seems to have dressed for the beach. Instead, they have brought chairs. Lightweight, upright wooden chairs, and I had always wondered at their stability on the sand, though there must have been a degree of relaxation as two of the crowd are reading newspapers.

Behind, on the skyline of the painting, are buildings I could see even now – or at least could make out in the dawn light. The obelisk erected to commemorate the town's designation as Royal Harbour (for its hospitality to King George IV). Castellated towers of what is now a fish-and-chip shop. The Georgian terraces on the East Cliff and the great chalk cliff itself.

Frith was one of the first Victorian painters of daily life. Hugely popular (the painting was purchased by Queen Victoria) and critically divisive (many thought it vulgar), his work rode a tide of change, that featured art about and for the masses. All those individual, unselfconscious, and intimate moments captured, plus the unlikely and strangely voyeuristic viewpoint – not at the edge of the water, but actually in it – had often made me wonder how much joy he would get from a modern iPhone camera.

I thought then how I'd like to walk the streets of London

with Frith's ghost, and I wondered whether John Lubbock could arrange an introduction – they must surely have known each other. If that wasn't an insane thought, it was almost certainly pushing it.

I was about level with the great rise of the East Cliff from Frith's painting. Its white face – blushing dimly pre-sunrise – looked natural. And it was. But 450,000 years ago, the beach I stood on was thirty metres under the ground, deep beneath a chalk ridge thirty miles in span that stretched all the way to today's France. This Weald-Artois ridge, as named, was knocked out by a massive glacial lake to the north that once sat between the coastlines of southeast England and France. A build-up of water caused a breach to the south and, in a catastrophic megaflood, the waters of the lake – in effect the North Sea – flowed across the ridge, scouring the land and then the layers of chalk beneath with its flow until the entire land bridge had been washed southwest into the Atlantic. The cliff was the 21st-century face of the breach, solid-seeming and studded with horizontal bands of flint. An iron-railinged promenade ran along the top, set with Victorian shelters offering fine views of the sea.

I turned north and followed the tide line in the direction of Broadstairs, hoping not to fill my pockets with too much flint. Washed straight from the chalk, the nodules here gradually lose their cortex as they're tumbled by wave action and range from chalk white (newish) to bruise grey (beautifully worn) through smooth iron-pan orange (occasional).

Lumpen balls, L-shaped thumbs and skinny fingers, their patterns repeated and as sculpture were hard to resist.

Here and there were pieces of sea glass. I bent down and turned a shard over in my fingers. It was modern bottle glass. Amber brown and softly frosted. But sometimes the tide on Ramsgate beach brings with it small pieces of fancy glass and patterned pottery. They seem no different from the broken bits found in garden borders and ploughed fields – which are standard survivors from domestic middens dumped on the land as fertiliser. But fine china and lead crystal have no place on a beach and their source is extraordinary. For they come from the sea – or more exactly a place beneath the sea called the Goodwin Sands.

The Goodwin Sands is a bank of dunes that lies to the south of Ramsgate, around four miles off the coast of Deal. Ten miles long, and twenty-five metres deep, the Sands sit on a solid platform of clay on chalk, and at low spring tide, these dunes of fine, evenly sorted grains solidify and – briefly – become dry land. Something like eighteen square miles of land. Photographs exist of jolly men and women of Kent playing cricket on the expanse, a tradition that began in 1824 at the instigation of Captain K. Martin, the Ramsgate harbourmaster. There are 'news' films of football matches, and later videos of hovercraft excursions. There seems to have been no end to the transient jollity on this ephemeral beach.

The Sands sit at the southern end of the North Sea in the Strait of Dover – probably the busiest shipping channel in the world for at least two millennia. On a map, the Sands' vague M shape looks like a ragged raven taking flight – or perhaps I just choose to see it that way, just as I see the Sands as a 'she'.

Way to the north lies the submerged landscape of Doggerland, flooded millennia ago in a series of tsunamis. Myth claims the Sands were once an isle themselves (Lomea, according to a 16th-century writer) but this is not true. They are, instead, a strange waltz of time, tide and sediment, with the undersea dunes fed and held captive by ebb and flood channels, all the while shifting with the waters' flow and losing breakaway banks to the shore.

Like all the best monsters, the Sands have an ambiguous relationship with the human world. They are known as the 'Ship Swallower' because more than two thousand vessels are thought to have been wrecked there, run aground in storms and sucked under as the turn of the tide transformed land to quicksand. From Bronze Age trading vessels to Second World War aircraft shot down over the Channel, few of them, their cargos, and their poor crew, were ever recovered.

'It must be that buried deep in the Goodwins, where the sand meets its chalk base, there are literally generations of shipwrecks compressed and "concreted" into a solid mass, holding an unbelievable treasure house of gold, silver, tin, jewellery, copper, brass and other valuables, probably forever inaccessible...' wrote her biographers Richard and Bridget Larn, in their book *Shipwrecks of the Goodwin Sands*.

But herein lies the ambiguity. Trinity Bay and the Downs, the deeper channels between the sands and the shore, have provided safe anchorage for millennia, with the Sands serving as a breakwater. So they have – according to the Larns – 'probably saved a thousand ships for every one destroyed'. It's argued that the Sands' actual name derives from Anglo-Saxon *gōd wine*, meaning 'good friend', not out of sailors' irony but respect for the place that offered shelter in the wildest winds.

When it's hot, it's hard to imagine being cold. When it's clear, it's hard to conjure a sea mist. When it's calm, it's hard to imagine a raging storm. The sea in front of me was like a millpond, blazing with the multicoloured light of a midwinter dawn. The air was chill and my hands were stiffening inside my gloves. Across the Channel I could see France, a dark rise against the orange band. My line of sight travelled right across the submerged Sands and I wondered at the drama and tragedy of it all, and whether there were ghosts out on the water, even now. I remembered a ship called the *Lady Lovibond*, from Folklore, *Myths and Legends of Britain – a Reader's Digest* bible of my childhood. There was a story of unrequited love and murder on board, and a tragic end to a wedding party as the ship sank into the Sands in 1748. The ship was said to reappear every fifty years, and in 1998 ghost-hunters filled the guesthouses of Deal to witness the marvel. But the ship did not appear. Newspapers reported that fact, but not the ghost-hunters' reaction. I felt sad for them. The whole tale was said to be a myth, and the *Lady Lovibond* is not on the extensive list of ships lost.

I unlaced my boots and slipped out of them, peeled off my socks, rolled up the legs of my jeans and walked two paces into the water. It washed my ankles and bit me. The cold sand felt like snow. I thought of all that death. The fortunes lost, the lives ruined and the hope sucked under. And the taunting shards of crockery and glassware spat back on the beach by the monster (who, acknowledged, saved more than she swallowed, but everyone loves a villain). Like grape seeds, the shards were. Like pieces of eggshell from a homemade cake or shattered stones in plum jam, the monster crunched and regurgitated the ships' table settings as the sands below shifted with time and tide, all the while keeping their treasures.

A friend, another veteran beach walker, has a magic eye for the shards and keeps an elegant, wide-mouthed bowl for her finds. Some of them are likely very old. Maybe last washed in the galley of some 18th-century brig. From the hand of the cabin boy who rinsed and stacked them to hers, across one, two or three hundred years. I pictured the spectral/corporeal exchange and wondered at the rare and random things that cross time.

The sun had yet to rise but it was close. I'd found a couple of rounded nodules, nicely weighted, and held one as I squinted at the horizon. My feet were numb in the water. Then the brightest point flamed and it was up. I hurled the flint into the sea, making a wish as it hit the waters and the sun's light left me blinking violet moons. A beach crow, browsing on the rocks with its partner, took off to the left of me, cawing.

I waded, woozy with swash and backwash, back to the beach and dried my feet on a bandana. Sand stuck but my boots felt warm. A slew of flint lay before me at the tide's edge and I couldn't get distracted, so I walked up closer to the cliff. Peach sunlight caught the tips of the ripples, casting mauve shadows. It was a Fauvist wonderland. I took a photo and picked up a round sponge and a fat flint finger.

The cliffs, parallel to the sea near the town, curved in and out the farther I walked, as if pressed out by a giant cookie cutter. In the first inlet I came to, the sand on the beach was washed so thin as to reveal the underlying rock. But the rock was chalk. It lay in some places flat like a pavement. In others it rose as high as my knees, like huge, holed cheeses

– a sculptural expanse of white, pitted and hollowed, dotted with flint and draped in seaweed, with sand and small pools in between. The pools were no more than scoops, but they hosted small crabs and winkles.

Sunrise caught the tiny ripples on the water. Everything glittered.

I trod over and around rocks and pools straight out of a storybook. There were no mermaids. No friendly sea dragons, but the beach had gone magic and there should have been. I looked back to the cliff base and some concrete blocks where a jolly sailor splicing rope, with a striped shirt and tarred pigtails, might have been sitting. There was no one there. Or maybe I just couldn't see him.

The chalk beach was becoming impassable, and I wandered up to the dry sand by the undercliff. I was reaching a small promontory and I hoped that in the next bay I would find what I was looking for. A low concrete esplanade lay up ahead, disappearing around the next curve, and I stopped on its steps for coffee from my flask and a cereal bar, squinting out at the sea in the raking light.

There were people now. 'Happy Christmas!' a jogger with a dog called out to me.

'And to you!' I called back, waving.

I shook out my cup, pocketed the cereal-bar wrapper and carried on along the beach. The cliffs were lower here and crows were calling in the trees along the top. I reckoned I was somewhere between Ramsgate and Dumpton Gap, so not far from where I needed to be. I bent down to pick up a small, rounded piece of flint. The shape had caught my eye – almost a heart but not quite. On one side it had the clear markings of an echinoid. It was only slightly water worn, so probably fairly fresh from the chalk. I dusted the sand off and put it in my bag, rounding the cliff to the next inlet.

I saw the first one because of the light. The sun was kissing the top of a large, solidly embedded blue-black flint nodule sticking out of a decaying chalk pavement. I wandered over, hopeful. It was about the shape and size of a small Christmas wreath, deliciously globular and tide-washed of its cortex. It had a solid core of cream chalk. This was what I was here for. I got out my tape to measure it and took a photo, tape for scale. Then I set up my phone on a piece of flint and took a photo with me for scale. My first paramoudra flint.

Also known as potstones, paramoudra flints are large and doughnut shaped – Victorian geologists say pear shaped. They rise around a hand's-depth, and occasionally up to a half a metre, from the surrounding chalk and sand, encircling a core of chalk. They're like hagstones for giants.

P-a-r-a-m-o-u-d-r-a. I love the name. I love the way it feels as I say it. I love the fact that the word appears in a poem about a geological lecture by the geology warlock William Buckland, who supposedly coined the term after his work in Ireland in 1817 (*padhramoudra* = ugly Paddy, not necessarily the kindest geological term). I love the engraving from Ole Worm's museum catalogue of 1655, featuring an unmistakable specimen. I also love that not everyone agrees on how they were formed.

FACT:
Paramoudra flints are found in horizontal clusters across beaches in northwest Europe, rising – slightly – above the chalk pavements in which they're embedded. They are also found in eroding cliff faces, piled one on top of the other, like great lithified spines.

THEORY 1: THE LAIR OF THE UNNAMED WORM
First the chalk formed on the seabed. There were enough nutrients to feed a community of crustaceans and large worms. The worms created a network of tube-like burrows deep below the soft chalk seabed.

The worms died and decayed. Sediment filled the burrows. Nodules of pyrite formed, and other iron-rich minerals drifted out from the sediment, clumped together and stuck to the walls. Sometimes the burrows passed through silica-rich areas of the sea floor, and when they did, this silica-rich water was drawn, through the chemistry of the sea floor active at the time, almost like a magnet, to form a ring around the outside of the burrow, and gradually hardened into flint. The paramoudra.

THEORY 2: THE REALM OF THE GIANT SPONGE

Charles Lyell, sometime student of Buckland's and fellow geological magician-priest, considered paramoudra flints to be lithified sponges that lived in great undersea communities.

And it's true the paramoudras do look like large sponges from sponge forests of today (and sponges were the primary source of the ocean silica that formed flint). The current theory is that they are barrel sponges: great hollow creatures (for sponges are classified as animals, not plants) with a glassy silica skeleton that can grow – while anchored to the seabed – as big as a small car.

I wanted to see the paramoudras for myself and now I had. There were around a dozen of them, spread across a small patch of sand. They varied in size – though not greatly – from forty to fifty centimetres across. Some were more ring-like than others. Some were close; others set apart. Large non-doughnut-shaped nodules lay here and there. Were they ex-rings, or were they never rings at all? All were set or encircled with chalk – in fact the chalk made up about 80 percent of the formation. The flint had sheared in places; inside it was a rich raven black. I wanted to call the whole mass lava-like, but actually it was more like blobs of black-and-white shaving foam.

The popular theory seems to be Theory 1. But having seen the paramoudras for myself, Theory 2 made much more sense to me. All sponge types of which we are aware lithified into flint. Their living skeletons were made of the very substance from which flint was formed. And if the silica-rich waters really were drawn to the tube burrows like some chemical magnet, then they would surely create a more solid sheath, and not a vertical string of spaced-out giant scrunchies?

Theory 2's explanation for the stacking seemed plausible, if a little sketchy, as I had to imagine myself under water on a cloudy seabed that sat on a shallow shelf to the north of the long-vanished Tethys Ocean – a great prehistoric ocean from the time of the dinosaurs. As the sponges grew, went the theory, their weight led them to collapse into the sediment, which in turn was rising, albeit at the glacial rate of 18-25mm per millennium. As one sponge sank, its offspring rose to take its place, and so on, in a cycle that lasted millions of years.

I stood back and took a photograph of the beach and the flinty-chalky, 85-million-year-old relicts. Nothing really dies in this universe, I thought, but then the sponges were very dead to themselves. I set up my phone to record the scene, with me in the picture for scale. The sunlight was still peach coloured on the flint-tops but it was turning. I posed.

'Do you want me to take a photo?' called a dog walker coming from the Broadstairs direction.

'Thanks, but I'm just the scale,' I called back, feeling a bit awkward. Then wondered if that made sense to her. She waved in acknowledgement, smiling, so it must have.

I shot the reef from all angles, and in close-up until the morning light turned yellow. It was now officially Christmas Day and the beach had ceased to be mine. There were seabirds everywhere. And quite a few people. Also, the tide was coming in. I turned back to the town, this time on the concrete walkway by the undercliff. I walked, glancing to the sea, at all the silhouetted birds I couldn't identify. The walk took no time. Before the pavilion, I cut down onto the sand. The original walkway, above me now, had been cut away for the last few hundred metres and replaced with a wide concrete platform, like a single storey of a multi-storey car park. It stood around five feet above the beach on pillars. Just inside, propped on the sand and leant deliberately against one of the pillars, was a small picture frame. It was side-on to me, but I could tell by the sun's glint that it had glass in it. I walked over and ducked beneath the platform. The frame held a photo of a dog. A black-and-white collie. The type that look like they are smiling. Below the picture was a tea light and a lighter. I picked up the frame and turned it over. There was no name. Nothing at all. I put it back carefully. Someone had loved the nameless dog enough to set up the shrine, I guessed because the dog had loved the beach. I undid the straps of my bag and groped for the fossil sponge I'd picked up earlier and placed it to the left of the frame. It was about the shape and size of a golf ball, and I hope he or she liked it out in the great beyond.

I passed the pavilion. Nothing stirred inside. The sun streamed in on tables set for Christmas lunch, gleaming on glasses, spoons and forks. I rounded its corner, stepped off the sand onto the concrete slipway and walked, past the seafood van, the bollards and the public toilets, into the harbour square.

The paved centre was empty but for two people. She was sitting on the pediment of the obelisk, he was pacing twitchily around the edge. They were both smoking cigarettes and laughing. And shielding their eyes against the sun. He was rake thin and dressed in chef's whites. She was in a grey sweatshirt and pale jeans, blonde hair scragged back in a ponytail. Across the road next to the castellated fish-and-chip shop, bar doors were folded open to the sun. I guessed this pair were on their first break of the morning.

'Happy Christmas,' the woman called to me and waved.

'Happy Christmas to you!' I called back.

She said something to the chef and they both laughed again.

I walked nearer and took a deep breath. Their smoke was lovely on the cold morning air.

'You know you've got one of the legs of your jeans half rolled up?' the woman said, pointing.

'Oh...' I looked down. 'I've been in the sea,' I said. 'Paddling.' I was oddly embarrassed, and bent to roll the material down.

'I love that,' she said. 'That's spiritual. You went paddling on Christmas Day.'

'Yes...' I didn't know what more to say. I didn't feel spiritual. I felt hungry and my feet ached.

'This is a really spiritual place. There are ley lines here,' she went on. 'I love it. I've been here four years. I love the beach. Specially the beach. And the sunrises. Specially them, too. Big skies.'

She turned to the sun, eyes closed. Face raised and arms to the side, palms out, like a praying Roman. High on the morning. Cigarette between her fingers.

I inhaled.

'Sometimes this is the most beautiful place in the world,' she sighed.

I looked across the beach again to the sea. I thought of the sponges that died, were turned to stone for millennia then surfaced in another world. I thought of the violence that brought the sea into being. I thought of the lost of the Goodwin Sands. The no-show jolly sailors and the mermaids. The dead dogs and the deader sea urchins. The water glittered. Gulls wheeled.

She was right.

'We've got to get back to work,' said the woman, shifting out of the moment and grinding the cigarette under her trainer, keeping hold of it for thoughtful disposal. 'Come on,' she said to me. 'D'you want a coffee? I'll make you one on the house.'

We crossed the road to the bar and walked inside. I sat on a barstool nearest the door, not wanting to get in their way, and enjoying the stillness after the perpetual motion of the beach. All the furniture, and the wooden floor, had been painted black, and dust motes floated in the light blasting through the windows. I scrolled through my paramoudra photos on my phone.

The woman – Linda, she told me, as she brought me a milky coffee in a glass mug – had come to Ramsgate for the music scene, she said, reeling off bands I'd never heard of and others I'd never imagined were still playing. How she got to talk to the guitarists and the singers and all the famous people who came to listen. She wasn't young, and if she had another story – the real story of how she came to be here – she wasn't telling it. The coffee was hot, and its steam joined the motes in the sunlight. I was grateful for it. Linda had put a biscuit on the saucer, and I dunked it.

There were a few more people about now, but not many. Most would be waking to their Christmas morning rituals – the general and the personal. Most kids would probably have been up for hours on this most thrilling morning of the year. So far, it had also been one of the most thrilling mornings of mine.

Linda was back behind the bar and busy, and the chef had disappeared into the kitchen. I needed to get back. But I suddenly wanted to give her something. Something for Christmas. I looked in my bag.

I found her changing a barrel of beer.

'It's Christmas Day and you're working,' I said. 'Here...a present from the beach.'

I handed her the flint with the sea urchin imprint. She took it and turned it around in her fingers. Then came round the public side of the bar and kissed me on both cheeks.

'I'll put it in my jar,' she said. 'My raven jar. I collect things like those birds do. Crows do it too.'

She went back round and reached up into a large glass jar on a shelf next to the whisky. It was about three-quarters full of broken china, glass and shells. There were a couple of pretty pebbles. She fished a shard of china out and handed it to me. It was pale cream coloured. I turned it over. It was crazed with traces of a blue pattern – part of a rim for sure, probably of a large plate. I could make out what looked like the top of a couple of sails and three jaunty blue pennants. Above was some lettering. There was a fancy L and a C. But the plate was worn so it was more likely an O. It was generous of her, and

I thanked her.

'A gift for a gift,' she said. 'I got it from the beach. I collect it, china and glass. And pretty stuff. Sometimes there's loads of it, sometimes there's none.'

I wondered if she knew where it came from. If she did, she didn't say. If she didn't, I wasn't going to tell her.

I thanked her, said goodbye and promised to call in next time I was down. It was a short walk back to my friend's house and I hurried, with the feeling I'd been away for ever, nodding at the dog walkers I passed who were now out in cheery force.

The fragment of shipwreck china Linda had given me was tucked in the small front pocket of my jeans. I doubted I would get a better present this Christmas; certainly not one so poignant. The lettering was likely part of a motto or even the ship's name, should the vessel have been fancy. An L and a C. Or an O. More likely an O. A ship was more probably named Lo... than Lc... .

I was nearly at my friend's front door when I realised what the letters on the shard might stand for.

chapter nine

INSECTAGEDDON

There are two small black-and-white photographs in a slate-leaved family album from 1959 that show my mother, then my father, in front of a wooden Viking ship. My mother is sitting on a grassy mound in front of the ship, which rests in a series of concrete cradles on a rectangular plinth, surrounded by a metal fence. She is wearing a cropped and waisted jumper, Capri pants and a pair of lace-up plimsolls. Her hair is wavy and jaw length and she is gazing off into the middle distance. She is beautiful.

My father stands in front of the fence, smiling directly at the camera. He looks confident and handsome in a hand-knitted cowl-neck jumper. It was the first holiday after their wedding, so a kind of honeymoon.

On the same page of the album is a black-and-white postcard of the ship, with deckled edges. It has small white lettering that reads *The Viking Ship, Pegwell Bay, Ramsgate.*

As a child I thought this Viking ship the most romantic thing possible. I did a school project on the Vikings featuring a drawn and coloured, horn-helmeted Viking and a horn-helmeted 'Vikingess' with braids. I drew the ship on the cover of my 'book', and on the inside. As far as I was concerned, that Viking ship was real.

And in a sense, it is. Named *Hugin* after one of Odin's ravens, the ship was built in Denmark in 1949, a gift from the Danes to the English to mark the 1,500th anniversary of Hengist and Horsa's invasion at Ebbsfleet. It was a reconstruction of a later ship, from the 9th century, though nobody seemed to mind this, and it was seaworthy, having been being sailed across the North Sea by fifty-three Danes to Viking Bay, Broadstairs. A silent British Pathé newsreel shows jubilant costumed Danes approaching the shore and a bay full of skinny Brits. There are soldiers on the beach, Prince George of Denmark and even a nun in the crowd. The bearded Danes were rewarded with letters from the postman, beer and a feast.

In 1965, the sailing part of the newsreel featured in a *Doctor Who* serial *The Time Meddler*, the Danes standing for 11th-century raiders.

Fifty-four years after the episode was screened, and seventy years after the ship's arrival, on a Saturday morning in June, I sat in the passenger seat of a Peugeot 206 convertible, a hot breeze shredding my hair as I rode with a friend west out of Ramsgate. We turned south on the Sandwich road, through Pegwell, past the Pegwell Bay Country Park and on past the Viking Ship sitting on its plinth in all its morning glory. I had time – just – to see that the dragon heads at the bow and stern had been painted parking-line yellow. And we were gone.

There were no clouds to temper the sun, and England felt like France, with empty roads and fields of barley, power lines and heat. The tarmac shimmered. We followed a sign to Sandwich, taking us – unexpectedly – off the road we needed and into a lattice of town centre streets, with crooked, jettied houses of timber, brick and chequered flint.

We stopped in the car park by a river fringed with willows to check our route. The place was Sunday-evening-murder-mystery perfect. A man on a mobility scooter pulled out of a side street and past the round towers of the town's 14th-century defences. He was wearing a primrose-yellow blazer and a Panama hat. The willows hissed. The breeze was warm. It was only ten in the morning and we were both turning pink in the heat.

Sandwich was once a port on the shore of a waterway, the Wantsum Channel, but now it is landlocked, the coast lying two-and-a-half miles to the east. The Wantsum strait once carved a half smile more than two miles across in a trough of chalk, from the southern bank of the Thames estuary to the eastern coast, where its waters flowed into the North Sea. Thanet – the most easterly tip of southern England – was an island.

The eastern mouth of the Wantsum Channel was protected from the open sea by a long shingle spit – the Stonar Bank – that extended from Walmer and Deal in the south for eleven miles to Cliffsend.

This skinny bank of flint washed south to north by the tides had a ruinous effect on the Wantsum Channel, as did a smaller bank in its mouth to the north near Reculver. Silt carried into the channel from the River Stour could no longer flow into the sea, and the strait, which had once carried tall ships from the Thames Estuary to the Channel, filled until it was barely navigable. Bold attempts were made to clear the silt, as an impassible channel would ruin the livelihood of the people of Sandwich, the more northerly port of Stonar, and farther. Danish raids in 1009 laid waste to so much of Kent that all clearance stopped, and by the time the people of Thanet and the mainland turned their attention back to the job, it was too late. Some profited, though. The monks of Minster blocked small rivulets and created land, growing crops in places that had once been water, but all they really did was to hasten the inevitable. By the 16th century the Wantsum Channel was a wide marsh.

Out of the town and across this marsh we drove, turning off at the next roundabout, travelling along B-roads and onto a hedge-lined country lane that bent around land parcels and long-vanished trees.

We came to a hedgeless run, bounded by wire fences and with rape fields either side. Small brown birds flew out of the crop and bobbed for a minute beside and in front of the car. It felt joyful and wild. We felt part of a magical world. Rupert Bear and Bill Badger on a flying sledge, towed along by bluebirds.

We came to the sign. It was a two-metre-high English Heritage panel and it read 'Richborough Roman Fort and Amphitheatre', with opening times and entrance fees below. What it didn't say, and should have, was what was on the website: 'Richborough is perhaps the most symbolically important of all Roman sites in Britain, witnessing both the beginning and almost the end of Roman rule.'

We were the third car in the large car park of this symbolically important site, and one of them must have belonged to the site manager. A path from the car park led close to the River Stour, where a hand-painted sign read 'Boat trips to Sandwich.' We could just glimpse the river through the trees that grew on a steep bank cutting sharply down to the water and railway track.

We paid in the shop and stepped out onto the site. Three vast flint walls lay to the north, south and west, casting sharp shadows on the grass. The eastern wall had fallen, undermined by the River Stour, although it was hard, on this sunny morning, to see how that could happen. The walls had clearly been thrown up by giants and looked mightier than time. While they were crumbling at the top and the ends and the corners were missing, they remained the biggest flint walls I ever recalled seeing. Even as I stood on the ground,

they gave me vertigo. The other visitors – a couple with a baby in a pushchair – were walking by the western wall. They looked like they'd been shrunk by a spell.

We were standing in clover, wondering where to start. Or I was. My friend, who had been looking around at the grass, had other thoughts.

'Where are all the insects?' he said, leaning down and poking a finger in the clover leaves. 'There really aren't any.'

And he was right. there weren't. We listened and watched for a bit. No bees, no beetles. One fly that I saw. My friend wandered off in the direction of the amphitheatre to see if he could find them. Though in my experience, insects found you.

Between and without the huge flint enclosure was the scarification of ex-walls, foundations, banks and ditches, green with chalk grass and short meadow flowers. Information boards at key points told the story of the site and by my interpretation it went like this:

Once upon a time in the Iron Age, the land we stood on was an island that lay off the shore at the southern end of the Wantsum Channel. On the island was a fortified farmstead, maybe an idyll of woodsmoke and log boats, sheep, drying fish and the slap of oars on water, though more likely a trading post. Maybe it had mooring and offered beds for the night.

The island would have been well known to the people of the southern and eastern lands. Everybody traded. Maybe many of the fancy Roman goods coveted by wealthy Britons were landed here. There's no evidence of this, but Roman culture had been trickling into southern Britain for centuries. So why not?

The island was named *Rutupiae*, possibly a Romanisation of a Celtic word that might have meant 'crossing point' or even 'mud'.

In AD 43 the military came. Roman soldiers, under orders from Emperor Claudius to take Britain for the empire, made the island their beachhead. Whatever it had been to the Iron Age Britons, it was now the Romans' land, and the soldiers dug two ditches parallel to the coast to protect themselves, their supplies and equipment from attack. An entrance in earth ramparts had a timber gate topped with a parapet. This gate would mark the beginning of Watling Street, the road the Romans built to run all the way to the port of London, and where, on its final cobbled stretch to St Paul's Cathedral, you can still buy fancy pies today.

Rutupiae became a military base. Gridded streets divided large wooden buildings for grain and supplies, and these gave way to shops as the base became a civilian town and import/export centre for the empire. This port grew into a larger seaside town with grand masonry buildings and landscaped gardens. It was the 'Gateway to Britain'. Around AD 150 this was celebrated with the construction of a massive cubic triumphal arch. With a height of twenty-five metres, it is thought to have been one of the largest arches – if not *the* largest – in the empire. (For comparison, Marble Arch in London is only fourteen metres high.) The arch's date – deduced from a coin found in the construction mortar – suggests it was built in honour of Antoninus Pius, adopted son and successor of Emperor Hadrian. Antoninus gave his name to the turf fortification of the Antonine Wall between the Firth of Clyde and the Firth of Forth – the northernmost frontier of the Roman Empire.

The marble that clad Antoninus's celebratory arch came from the quarries of Carrara in today's Northern Italy. It was decorated with statues made of bronze. Hotels, shops, food stalls, a bathhouse, an entertainment complex in the form of an amphitheatre; the port of *Rutupiae* spanned twenty-one hectares and was – briefly – the greatest town in Britannia. A bright Disneyland of a place, lent by the Romans every pigment available, as the reconstruction drawings attested in their gaudy glory.

I stood in the centre where the arch had been, its cross shape marked out in beige gravel. I imagined the ship that carried the marble – and the masons, possibly – sailing up to *Rutupiae*'s docks. Unloading the stone in straw-packed crates on the cool foreign shore. Lashing slender scaffold poles with rope and raising the quadrifrons – or four-fronted – structure to its magnificent height; branding England as their own. How glorious it would have looked, the port raised on Mud Island, on a day like today! I imagined it in November drizzle, and its glory faded a little.

Time, tides and Saxon raiders saw a gradual end to the port's prosperity, and in the late 3rd century the military took charge again. The empire was under threat at its fringes. Pirates sailed the North Sea and internal strife rippled through its lands. *Rutupiae* was recreated as a military fort – one of a string around the southern and eastern coast of Britain. The forts were not flamboyant, they were not for civilians, and the world – especially Roman politics – had changed. Two hundred years after it was erected, the triumphal arch that was meant as a stamp on the land was torn down. It had lasted just eight generations. London's Wellington Arch, built in 1826, has almost out-endured it.

This new phase of the settlement saw walls built with corner bastions ten metres high and nearly three-and-a-half metres wide at the base. Their cores were made of flint and concrete and they were faced with fancy square-cut limestone, sandstone and ironstone, quarried nearby. Six bonding courses were made of red Roman tile.

I wandered over to one of the crumbling sections and reached up and touched an irregular white block that was a little higher than my head. It was silky. Like finely milled face powder. Chalk?

My finger held a light slick and I rubbed it in a line across my cheek to my ear. The wall facing had fallen away just above it and I could see the core. Huge flint cobbles bigger than my head set in a matrix of porridgy Roman concrete.

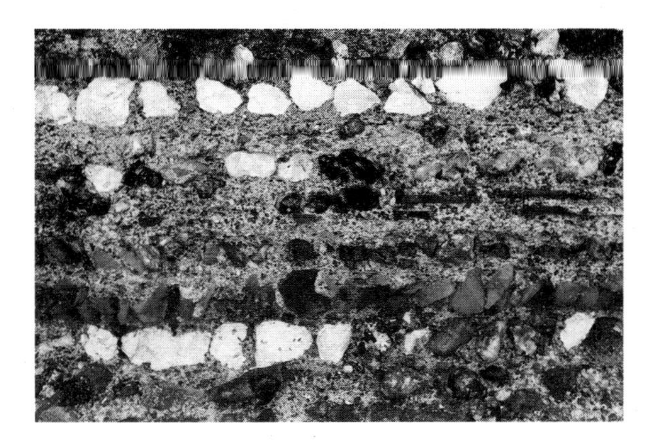

Bright stone flecks stood out of the concrete mix like chopped raisins and cranberries. The cobbles were sea smoothed and mostly oval. They were covetable. They had been gathered – from the Stonar shingle, geologists had proved – using baskets, boats and sweat.

The walls' construction would have turned the town, already broken by politics and diminished by raids, into a filthy, stinking building site, before its next performance as a fort.

The hot breeze held the faintest notes of grass, hay and rape flower. Sweat and shit and spit and smoke, burnt meat and stew and fish sauce. Stale wine. Tanned leather. Filthy jokes and curses, laughter. Triumph, pride, ambition and sorrow. All of it gone. Two millennia separated us from the first soldiers who dug the ditches and ramparts with their military-issue shovels; only slightly less from the masons who built the Shore Fort's defensive walls. There was nothing left but stone and scars.

My friend was back from his insect quest. 'I wonder if the pesticides they're using here have done for them?' he said. We looked across to the fields that joined the site. 'Perhaps this is it,' he went on. 'Insectageddon. How long would that give us? Because there aren't many birds either. About a year until the crops start failing? Maybe two?'

I thought about it. The Romans who built Richborough clearly thought it would never end. But everything ends. Thanet as an island, the great Wantsum Channel, the civilisation of Rome. A slow roil of land, water and life. Insectageddon would be the biggest end of all. We would lie, all of us, as a geological layer above the clay, the flint and the chalk.

I did another turn of the western wall. Wiry grass thrupped against my shoes as I walked. I took a photograph of some lovely daisies by an outer ditch, then I stopped by the last information board, which told of the construction of the walls and featured the white chalk blocks. Except they weren't chalk. They were the smashed and burnt remains of the marble arch, reduced to rubble and used as hardcore in the walls' fill. I looked guiltily at my finger. There was no trace left of the smear.

My friend was sitting on a limestone block by the trees on the river side of the site. The sun was burning now.

'How does it make you feel?' I asked. 'This...' I gestured to the landscape in its last iteration as art installation. Nothing looked less like 'significant Roman site'.

'I think it's all wonderful,' he said.

'Do you?' I asked.

But he meant it. 'I think of all the lives and loves and fights and dramas,' he went on. 'We don't know any of the names or the fates of the people who lived here for all those generations. What can we do? We've none of us got long. For myself, I'm just happy to be a passing visitor in this world.'

We drove back, both of us happy tourists in East Kent in a time designated the 21st century by our civilisation, on a sunny day on planet Earth.

After lunch we stepped out of the umbrella shade of the restaurant terrace and onto the hot harbour cobbles. The specials menu outside was written in chalk. Not bought chalk, but the worn edge of an oval pebble from the beach. Vegetarian lasagne, moussaka and sea bass written with an 85-million-year-old marker.

We were admiring the boats with first-time friends to Ramsgate when a butterfly came by. It flew around us while we took photos: all of us leaning on the harbour rail, smiling, sunglasses on. The butterfly persisted. It was a red admiral and quite beautiful. It fluttered close to me for a few moments then landed on the cuff of my sleeve. I slowly extended my arm in front of me to photograph it, focusing one-handed as best I could. We were close to the centre of town, but the profusion of wooden planters by the shops, and boats with flowering tubs on the decks, made it a reasonable habitat for a butterfly, I supposed. And as an adult butterfly's diet is

mostly rotten fruit and bird droppings, this was probably a much better place to live than the rural site of Richborough. It was almost certainly sitting on my sleeve because it wanted the salt on my skin. I saw it unfurl its tongue a couple of times. Then it flew off. My friend airdropped a handful of photos to me that I couldn't even see in the sun's glare, and we walked off into the afternoon.

Six months afterwards, the photobook arrived. One of those made from social-media pictures scraped and arranged by algorithm. Inside, among the walks, digs, friends and family parties, were two pages showing the Richborough morning. I'm standing, tall as ant, next to a gargantuan flint wall, wearing a red silk dress and red-rimmed sunglasses. You can just about see I'm smiling. There's a close-up of some flint cobbles set in Roman concrete. There's a huge water-worn nodule in another close-up, by my feet. I doubt I could lift it. There's a picture of my friend, sitting on a stone block to the east, close to the fallen wall. He's wearing a baseball cap, polo shirt, and chinos. He's looking back at the site to the southwest, laughing. But the photo the algorithm has chosen as its hero picture, full length on the right-hand side of the page, is the one he took of me beside the harbour. I'm standing by the railings, bright boats in the sunshine behind me, the harbour arm curving round in the distance. My left arm is outstretched and my knees oddly bent, my right hand holding my phone like a claw, as close to my nose as possible, thumb over the screen to focus. My hair is falling to my shoulders in rats' tails, bunched only at the top by my sunglasses which are worn like a hairband. My head looks too big for my body and my nose, from the angle, is swoopingly long. I am entirely foreshortened, but the butterfly on my sleeve looks good. And there was something else. I looked at the picture closer. A white blob – more a streak – runs across my cheekbone to my ear. It isn't a flaw in the photograph or dirt on the lens. It is very real – the forgotten smear of Carrara marble that my lunch companions had not even thought to comment on. The dust of the once-triumphal arch and glory of an empire was later, doubtless, smudged by a careless hand into the 21st-century ether or washed down the shower drain of a future the Romans could never have imagined.

chapter ten

THE FIELD OF STONE COFFINS

It was early in the morning on an early autumn day, and I was standing in thick-dewed sunshine on a path I didn't know, in a world that belonged to me and the birds.

'It's fine. Really. Just here is great!' I'd insisted half an hour before, when my cab driver, whose name was Arben, had a crisis of conscience about letting me out halfway down a steep, hedge-sided lane beyond a church in a place he said was 'nowhere'.

'I'm here to photograph wildlife,' I lied. 'The jays.'

It's a line that seems to satisfy most people. I had a decent enough camera with me and could have been telling the truth.

'How will you get back?' he asked.

'Walk,' I said.

'Noooo,' he said. 'Too far.'

'Or I might call you,' I relented. 'If I get too tired. I'll probably do that. Yes. Why not? I'll call.'

With that, he let me pay him, and I watched him reverse

into a farm gateway and drive away. He was Kosovan, and I suspected his worry had nothing to do with me. He had just had what was possibly his first attack of Downland panic. I could see why. The nature here was really thick. Like tree pollen, it caught in your throat.

I started down Church Lane, a tarmacked hollow way sunk into the chalk. Cars had scored the lane sides white and there was nowhere to throw myself should a vehicle come speeding by. And the lane was twisty. So I hurried to the bottom. In fact I ran, until I came to the meeting of Blackness and Jackass Lanes. The names made me smile. Pan was abroad here, probably sitting invisibly on a stile, laughing to himself.

To the north were some log steps and, according to my phone, the path I needed. It made up the fourth arm of the crossroads. It was steep and it climbed unrelentingly between two fences. Trees from the chestnut coppice on the left had dropped prickly seed carapaces all over the path and they mixed with fallen damsons from a bush on the other side. They were pretty and I stepped around them so as not to crush them underfoot.

I had kept my promise to myself, and I was heading off the chalk to Keston. More precisely Keston Common, as I was already close to the place that had given the original hamlet its name. Meantime, I was on familiar geology.

Somewhere, in the large garden of a private home on the hillside I was climbing (maybe even the house of the damsons), were Roman tombs. I had visited the site one

September many years ago on an open day. It was all sunshine and specimen trees, with a tea tent for charity. There was even cake.

'...Keston, originally Kyst-staning, the field of stone coffins...' wrote John Lubbock in his book *Pre-Historic Times*, referring to this field. The site was noted by antiquarians in 1815 and first excavated in 1828, when a number of stone coffins were found. They believed it to be the lost city of Noviomagus – documented in the Roman travel guide *The Antonine Itineraries* as being ten miles from London. (It was not Noviomagus, which probably probably lies a couple of miles to the east beneath modern West Wickham.)

Like most archaeological sites, there was beauty in Keston's curated remains, and here they were a fine sculptural arrangement of flint, accompanied by one – broken – stone coffin, on a lovely terrace slope with fine views north, east and south.

The walls were a simple, single note from a complicated tune of deep time, and I tried to play that tune in my head, standing there politely and alone, looking at a photocopied handout and drinking tea from the third-best china with the obligatory – perfectly baked – bit of cake.

The main circular tomb was almost nine metres across and today stood just a metre high. The flint was coursed, and long nodules had been chosen, laid slantwise, each course slanting in opposition to the first. It was the only time I'd seen flint herringbone, and it wasn't even for show.

The walls were just less than a metre thick. The inside was filled to its height with earth and grassed. It might make a fine garden feature, but it looked nothing like its original self.

The tomb was excavated, together with its rectangular neighbour and much of the surrounding area, beginning in 1967, after a couple of previous attempts had failed to do the job to the satisfaction of Bromley Council.

These excavations had been thorough, and this is the story they told:

The field was settled from 600 BC in the middle Iron Age, as post holes, potsherds and coins showed. Then the Romans – or Romanised Britons – came and built a farmstead here around AD 50. Slag found in pits shows they worked iron and bronze on a small scale, and they made pottery – kitchen and fine tableware to rival continental imports. They lined pits with clay for water and grain storage. They dug pits for rubbish and threw in their dead animals.

And they dug a ritual shaft in the chalk nearly five metres deep, in which they threw dogs, sheep, pigs, oxen and horses. The archaeologists counted fifty-seven animals in all, from their bones. The shaft was a squat bottle shape, with a wide chamber narrowing to a small neck near the top, with sides that appeared to have been worn smooth rather than tooled. (A similar shaft in Deal, about eighty miles east on the coast, contained a chalk figurine, the Deal Man, reckoned to have fallen from a side niche into the skeletal mire. Eighteen centimetres tall, the figurine was decanter-shaped with a flat, spooniform head, a fine nose, close-set eyes, a slit for a mouth and an utterly unfathomable expression – probably that of its maker.)

Shafts like this were common in Romano-British England and are thought to have been dedicated to an earth deity. On the floor of the Keston shaft were laid glass and pottery vessels, and a boot. These were the dedicatory offerings before everything piled on top. It was the first – and largest – of eight shafts found across the site, and while this one was reckoned to be in use for thirty to fifty years, their construction spanned nearly three centuries.

In AD 160 came timber buildings, and fifty years later a villa complex built just to the south of the tombs, complete with painted ceilings and a bath fed by piped water from springs to the north.

The circular tomb was a tower tomb – reckoned to be

between five and seven metres high when built, rendered in *opus signinum* – waterproof lime and crushed terracotta, a common flooring in many Roman houses – and painted red. The tomb itself was earth filled and had burials in the top, six flint buttresses keeping the whole from collapsing outward in an explosion of flint, clay, chalk and bone.

Tower tombs were known across the empire, mostly rectangular or square and made of brick or stone. This one owed its low drum shape to the flint, as flint doesn't do corners.

A rectangular tomb adjoining the tower came afterwards. Inside, in a shaft sealed by chalk blocks and tegulae – terracotta roof tiles – was a small lead cremation casket containing the burnt remains of an adult, a pig and a bird (possibly a chicken), the animals serving as an offering, or maybe food for the incumbent.

The tomb group was on higher ground than the villa and would have made a distinctive family vault, the great red tower rising two house-storeys high. In all, there were fourteen burials in and around the edge of the tombs, many of them children. Of the cremated adults, two were placed in pottery vessels with their hobnailed boots or sandals beside them.

The villa was remodelled around AD 300, and at about the same time the tower cemetery was abandoned. The new villa had a new cemetery to the northwest, while to the southeast a family member was interred in a single stone coffin.

One hundred years later the villa itself was left to the

elements, and the events that precipitated this abandonment cannot have been good for anyone. I remember the 'dark earth' mantra from a back-room class at the Museum of London: 'There was no continuity of occupation. Okay? Got that? No. Continuity. At all.' We scored it in biro in our notebooks, and I decorated the statement with spiky doodles, not really extrapolating the fact into the horror it must have been. The theory, borne out by archaeological science, was that with the fall of the empire, the Romans left, and the Britons they left behind eschewed the Romans' fancy ways. The Romans' heated homes and mosaic-floored public buildings were left to nature and a slow build-up of black loess – windblown dust, grit, seeds and decayed weeds – creating a small layer clearly visible as a horizontal line in the stratigraphy: the layers that accumulate on the ground over time.

As if to emphasise the point, Saxons erected a small sunken hut thirty metres from the dead villa, just half a century later. (It was these Saxons who etymologists believe gave Keston its name: 'boundary stone [stān] of a man called Cyssi' appeared in an Anglo-Saxon charter of 862. It's plausible, but we will never really know, and I choose to believe John Lubbock. Of course I do.)

What did the Romans call it, this place of the well-appointed valley farm on the London road? I thought of the newly painted red tower against green grass in the sunshine. I thought of how it would look on a frosty morning, and even in the snow. The Romans must have used a lot of ladders to get up their towers and down into their shafts, climbing in hard-leather hobnail boots.

I also thought of Pan, sitting on top of the tomb in the moonlight, playing a flute on a summer's evening. Or down the deity shaft waiting for gifts. Eating the votive food and trying a single Roman boot for size...

At the top of the steep path I was walking were more steps that took me up to the main road. It was fast and noisy with cars, even this early on a Saturday. I took the narrow, tarmacked footpath that curved north, past high brick walls and fences of grand houses, and I could just see through their gates into their landscaped hillside gardens that the Romans farmed. In minutes – though with the cars it seemed longer – I had crossed the road and was on the common.

And in another world.

Ferns grew by the path, and in places the gorse was still in flower. Somewhere under the road the geology had changed. I had stepped from the chalk to a gravelly mix officially designated the Blackheath Member (for many years called – more descriptively – the Blackheath Pebble Beds). These were the gravels of a relict estuary. A prehistoric braided (or multi-channelled) river – much like today's Amazon – had dumped tidal bars of pebbles and sand across the chalk, and in great washes at the river mouth. Mill Hill, above the Field of the Ghost River a mile-and-a-half away, was one such tidal bar, and there were pebble scatters all over the woods and even in my garden. Here, the gravels were widespread and deep. They likely hadn't moved much since the river abandoned them around 55 million years ago.

The path was worn to the raw geology of small pebbles and sand. It looked like a beach and in a way it was. There was a scattering of pink-to-red, similarly sized stones, and a few of quartz, too, but most were black, shiny and made of flint. I picked one up. It was small – the size of my thumb pad – slightly flattened, black and glossy as a crow, with tiny crescentic moons across its surface, the shape of a fingernail pressed in clay. These moons were 'chatter marks', the scars left by tens of millions of strikes with other stones in a watery environment like a beach or undersea bank. The pebbles are what became of great flint nodules that got washed into water. The element took them and broke them, then tumbled them against each other, transforming them into smooth aquadynamic pieces.

How the pebbles got from under the sea to the river, I wasn't entirely sure. Usually it's river to sea. Maybe the sea subsided, became land, and then the river came, only to become land again – the land where I was standing now. I suspected a slice of geology was missing, either from my knowledge or from the official record. Sometimes I think it's amazing that people have anything figured out at all.

Unlike the chalk and the clay, the land these gravels made was poor for crops and so, through its nature, luck and happenstance, it had been preserved until modern times as common land or heath. Everything here – especially the grass – was spiky. There were a few broadleaf trees such as oak, but most were pine, and of a twisted, unfamiliar kind. Even the trees' faces – when they had them – were different from those I knew on the chalk. It was beautiful, but not in the soft, careless way of the Downs. It was a prickly beauty, in acid shades of green and gold in the morning sun.

I came to a clearing that seemed dedicated to a tree. A low, warty pine that had gravitas*. A crow was browsing nearby. I threw it a couple of nuts, but it was obviously unused to taking food from strangers and it flew off. I hoped it would find the nuts later. Or maybe a squirrel would.

The path sloped down through some trees, which thinned to grass banks before opening out onto the shore of a small lake – the largest of the ponds. Trees backed the water on the far side, where three fishermen sat in the sunshine. They were motionless. I couldn't see, but I could imagine them asleep.

I sat on a wet wooden bench, the sun on my back, equally still. Some ducks slid from the bank into the water and drifted towards me. Another flew in and landed, scattering light from the water's surface. A contrail was manifesting in the sky, high above the pond, creating a white vertical line

* *Days later I scrolled through my photos of the morning. There was the warty pine. The picture before showed the tree again from a distance, where it appeared at the left of the clearing, bathed in golden sun, arms stretched up, svelte as the nymph it clearly was.*

through the blue. I photographed it, and its pond reflection. It looked like a soft crack in reality.

I didn't remember it like this at all. For what seemed like my whole childhood, but was probably only a couple of years, my mother or one of her friends would bring us here to throw sticks in the water. The trees were tall, we always wore Wellingtons and duffle coats, and it was always grey. But to my smaller self, it was the most thrilling place I had ever been.

My mother learnt to drive when I was six, and I pestered her to take me to Keston Ponds as soon as she passed her test. And she did. It was a glorious day, that day she picked me up from school in her Austin A30 and drove the lanes and the fast road to a muddy kerbside where she parked. I think we walked around one of the ponds and threw more sticks into the water. My memory of it, beyond sliding around on the red vinyl seats with my nursery-school-age brother in the back, is vague, but I know it felt like a milestone, heralding a world of freedom and adventure in far-flung places.

There were four ponds at Keston, three of them stepped and in a chain, each a little higher than the next. What I didn't know then was that they were not natural. A map drawn in 1790 shows no bodies of water at all. Keston Common used to belong to the estate of Holwood House, and the ponds were created to supply the house with water. What is amazing about this is that Holwood House, fifty metres to the southeast of the ponds, was also fifty metres above them. The hydraulic pump that drew the water up the one-in-one pipe has gone, but the brick-and-flint cottage that housed it still stands on the Westerham road.

The two most southerly ponds had been originally dug in the late 18th or early 19th century for gravel extraction. The pits were filled by a spring close by and were almost certainly lined with clay to keep them watertight. They quickly became a draw for sport and leisure. Keston Swimming Club had a New Year's meet before the First World War, there was skating when the ponds froze, and year-round fishing – boys used to cycle to the ponds with their fishing rods strapped to the handlebars.

My mother was brought here on outings as a child. Two buses from Elmers End. It was the place for South London people to come in an era when leisure was a day out with sandwiches, and an ice cream for the lucky. My dad spent his childhood here and at nearby Hayes Common, running wild. At Hayes he built a raft out of oil drums and floated it out into the water of a similar pond, where it sank with his friend Gerald Drake on board. My father and godfather ran away, as they told it, laughing in terror all the way home at the memory of Gerald surfacing above the sunken craft, still wearing his posh-school cap. They abandoned him to swim to the shore, then to run home himself in a uniform wet and covered with pond slime.

'But you left him and he could have drowned!' I would always say as a child, recalling stories of dog-saving heroism and similar derring-do from girls' annuals. 'But he didn't,' shrugged my father, laughing again at the replay of a decades-old scrape. The water gods got Gerald in the end. He died in a speedboat accident on a lake in Atlanta, Georgia, forty years after. I don't think my dad ever got over it.

The hill of Holwood House was a presence on Keston Common, though not a dominant one because the trees around the pond largely masked it from view. Old sketches from the 19th century show the place differently, almost as open fields, with the spring in the foreground and the small trickle of a stream. The house took its name from the hill and is recorded as being the property of Captain Richard Pearch in 1673. It was bequeathed or sold many times in the next century, until 1785, when it came into the possession of the twenty-six-year-old prime minister, William Pitt the Younger.

But Holwood was known by another name: Caesar's Camp. This was the place – reputedly – where Julius Caesar stayed with his soldiers on their march up to the Thames. Edward Halstead, in *The History and Topographical Survey of the County of Kent, Volume 2* (1797), was unsure:

> '...*it must have had great additions since from time to time, to bring it to that state of strength and magnitude which its remains now point out, for it is not probable that Cæ either had time to cast up such a work, or that he would not have mentioned so considerable a one in his Commentaries.*'

Halstead was right to have a doubt. Holwood is an Iron Age hillfort rising to 152 metres and covering eighteen hectares, built by a Kentish tribe around 200 BC. It was first surveyed and drawn in 1790 by two gentlemen from the Society of Antiquaries, Thomas Milne and James Basire, five years after Pitt's purchase. Shortly afterwards, wrote C. Roach Smith in the 1880 *Transactions of the Kent Archaeological Society*, 'the ramparts where the house stands, and also those on the north and south, were sacrificed to the taste of the landscape gardener and levelled, the entire area being converted into pleasure gardens'.

The western and part of the northern ramparts remain, with two banks and ditches and, on the western side, a third exterior bank in fragments. The banks were raised three metres above the land's surface, the ditches four-and-a-half metres deep. A natural escarpment to the southeastern side completed the original defences. On maps, the hillfort area looks like a freehand drawing of an egg. On Google Earth, it's invisible.

To me, this hillfort remained as far flung as the places I dreamt of when I was young. It was a concept, rather like Azerbaijan or Ulan Bator. I had never seen it, and probably never would, because it was on private land, but I had read about it and seen photos.

Hillforts took organisation and people, enough to sling the earth from the ditches to the ramparts with wooden spades or cow scapulae millions of times over. They were a statement of power and a place of safety in times of flux. They represented civic planning and control. They encircled houses, streets, metal-workings, granaries and animal pens. Sometimes they were seasonally inhabited, sometimes they were only used when people were under threat. Sometimes they were used year round and grew into towns. Sometimes there is no trace of human activity inside them at all.

Holwood's ramparts were excavated in 1956-7, broken

sherds of Iron Age pottery providing the date of its construction: 200 years before the Romans came. The archaeologists also found a scattering of Mesolithic flint. Hunter-gatherers were here, maybe 7,000 years before the Romans.

There is a story in the ramparts, beyond the fact of their existence. Not that you'd know it from 21st-century photographs, which show the ditches with wooded banks of beech, birch, ivy, oak, bramble and litter. It's clearly hard for the few fortunate visitors to make a good picture of them, and harder from those pictures to feel the energy and effort it took to build. The honesty of the excavation report is refreshing:

> *'The fort, with its multiple defences, represents a very large job of earth-moving for people without mechanical aids. Either it was done by a very large labour force, or it took a very long time.'*

And first, they had to fell the trees. Pollen remains showed that the hill was originally a dense oak forest, and likely had been for centuries.

The inner ramparts were built in three stages. First a mound. Then a higher mound with a flattened top. Then they raised the back of the mound and made a flattened top, two-and-a-half metres wide. This walkway on the ramparts gave a view to the London basin in the north, when London was just a settlement on a river, and across the wooded rises of the Kent Weald to the south.

There were six gates, and the main entrance to the fort was only 200 metres from where I was sitting on the bench by the pond. It was sited in a small natural valley leading up the hill. This entrance cut through the ramparts at right angles. Excavation photographs show that the gateway, which was four and-a-half metres wide, was revetted – shored up on either side – with flint nodules, carried here onto the gravel and sand of the fort itself. Burnt timbers were laced into the wall of the gate's north inturn, together with fragmented flint. At some point there was a massive conflagration, enough to burn the entire gate and shatter the stone. They had no defence against fire beyond leather buckets of water and sand.

I wondered if the fire was the last event in the fort's life, but on current evidence it's impossible to tell. In 1962, a drainage ditch was dug across the whole interior of the fort, north to south. Where one would have expected traces of the junk of everyday life – scraps of the lost, discarded and forgotten – the excavators found nothing. Not pits, postholes, flints or fragmented pottery. Holwood is a silent fort.

It also has a Saxon name. I wondered what its builders called it.

I didn't want to leave the bench by the pond. I wanted to sit in the sunshine and watch the world mirrored in the water for ever. I wanted to see if the fishermen caught a fish. Or even moved. There was birdsong, and the soft, repetitive *ssssushhhh* of cars on the road behind the trees on the far bank. Even the ducks had gone quiet.

But there was more to see, so I picked up my bag and walked the pebble path until I'd left the pond, and the fading reality crack, behind.

Within minutes and just a few metres, I was in a wide-bottomed, steep-sided hollow in the woods, approaching the spring that started it all. It goes by the name of Caesar's Well and is the source of the Ravensbourne, a river that flows north, all the way to Deptford Creek and the Thames.

The myth of it went like this: when Julius Caesar's army marched through Kent (and set up camp at nearby Holwood), the men were thirsty. They saw a raven fly down into the woods to drink and discovered the spring and the river.

The myth contains no shred of truth. The river's name might be Middle English for 'Rendel's stream', or Saxon for 'fast-flowing/stagnant stream', in a curious contradiction of translation. It might be none of those things, as rivers' early names are often lost. What's sure is that it had nothing to do with a raven.

I had no memory of the well from childhood and no expectation of what I'd find, but that didn't make it any less surprising. At the centre of the hollow lay a brick water feature, the kind of architectural conceit one might see in the forecourt of an eighties provincial shopping mall. The first part comprised a circular pool, about two metres in diameter, built of grey vitrified brick. The pool had a second brick

circle inside it filled with water, which was still, like a mirror. I stepped down into the ring, leant over and looked into its shallows. A woman-with-hair shape looked back.

From the outer circle, a brick-lined conduit flared out then back in again, feeding water to the north, out of the hollow and into the top pond. The water itself sluiced out of a slot at the upper part of the conduit, which was spanned by a red-brick platform. The platform covered the water's rise, presumably to safeguard the flow out of the ground. The geometry of the construct was as hard to fathom as it is to describe. Coffin-shaped with a round head? Or maybe a modernist goddess.

I walked around the pool a few times. I crossed the brick platform. I stood on it and looked in the direction of the lake. Then I looked down beyond my boots and watched the water flow softly out of the slot. I stood there for much longer than seemed reasonable. I took a hundred photos. I wanted to curtsey. I wanted to give a gift. I wanted to take all my clothes off and immerse myself in the source waters of the Ravensbourne – pointlessly, as the water would have barely lapped my ankles.

The myth also said the waters had 'healing properties', but don't all wells?

The hillfort builders wouldn't have been the first collective to drink here. The prehistory of the well likely stretched back 7,000 years as southern England came into its current geological being and the water table of the Downs – fed by rainfall – achieved a steady level.

Seven millennia of people passing, stooping and filling their water bottles or skins. And bathing. In the early 19th century, there was even a tree-fringed bathing house, attracting parties who wished to benefit from the spring's medicinal qualities. Or just to wash. I was grateful for the solitude.

A couple came down from the eastern path, a pretty blond dog bounding in front of them. He leapt into the middle of the pool and jumped around in the water for a few seconds before heading for the conduit and lake.

'Sorry about that!' said the man. I'd been standing idly with my camera on a monopod. I probably looked like I had more purpose than I actually did.

'Oh, that's okay,' I said.

'It's his birthday,' the man explained, then yelled 'Nooooo!' in fond exasperation as the hound got into the river, becoming half dog, half mud.

'Perfect birthday – for a dog,' I said.

The man laughed. 'He'll need a bath now.'

They headed down in the direction of the ponds and I was alone again. The ripples in the pool had stilled and I crouched down beside it and let my fingers drift in the cold water. I leant down and looked again. Trees rose above the woman-with-hair reflection. They made horns. If I stared into the pool any longer, I would see Pan staring back at me.

So I left, heading south from the hollow up a set of large, slippery log-revetted steps, across the car park and into the woods on the other side. The path was wide and covered in leaves. It sloped gently upward, but not as steeply as the banks on either side of me, which rose with every pace. Within about twenty paces they were as high as a two-storey house, and within a minute of leaving the car park I was walking in a great hollow. A dad with kids carrying sticks overtook me, the kids running ahead, up and down the steep ditch sides. It looked like a fantasy film set. I reckoned I was about level with the hillfort ramparts, somewhere up to my left, though this great ditch wasn't part of it. It was dug for gravel when the road was built. Or some such time.

When has there not been the most insatiable appetite for gravel?

There was a small path to the left that led to the top of the bank, and I took it. It was steep, with its own tiny banks covered in spongy green moss that sprouted the odd fern.

When I got to the top, the path disappeared and I found myself standing in a tangle of wild growth at the roadside, with fast cars passing just a metre from me. There was no pavement. I persevered, knowing I shouldn't, and had to stop three times to unpick my hair from low branches and a bramble. I had to walk close to the road where it was slightly clearer, in the netherworld, known only to booted hi-vized council workers clearing litter. They must have been here recently, as there was none.

I passed a small tree and saw a shiny coppery thing. On a tiny stub of a branch, at about chin level, someone had hung a bracelet. It was multiple-banded. Small copper leaves hung from one of the bands. Under its plating, it was just beginning to rust. It clearly wasn't old, but its age was hard to guess. I could see it on the wrist of a Roman girl as easily

as on a tattooed crop-topped mall-walker. The tree liked it, I was certain of it. The bracelet had been thrown from a car window, I reasoned, and had been found and hung there by the clean-up people. I Liked my own story and held on to it, as alternatives weren't an option. I was starting to feel foolish for not doubling back, but worse, I was feeling the first soft, suffocating waves of panic.

After what must only have been a few minutes, the footpath appeared, and I gratefully broke out of the common woods and onto the path. The road was fast and even louder than before. If I carried on for five minutes I could get to Keston Church, close to the road where I started, and make a plan.

I was two-and-a-quarter miles home as the crow flies. Farther by the paths. The day was lovely still, and I had coffee and biscuits in my bag. Though there was a part of me that didn't want to dim the foreignness of this landscape with the familiar by walking the land I knew.

Keston Parish Church is a low flint building with stone corners, at the confluence of five roads and a footpath. I opened the gate and went into the churchyard. The grass was patchy and overgrown, and the paths looked as if they were used more by foxes than visitors. The noise from the cars was intrusive, distracting from the gravestones. The church had the air of a handsome lady of great age in full makeup, but the fringes of the churchyard were thick with wild, late-summer growth.

I checked my phone. The church had no dedication and

no founding date. Bomb damage in the Second World War revealed a wall under the chancel likely to have belonged to a church that stood here before the Norman Conquest, and below that an earlier wall that (according to Historic England) overlay Romano-British chalk-dug graves from 300-400 AD. These graves may actually be later Saxon but, Roman or Saxon, it's interesting that the builders should set their altar on the dead.

From the outside, the church and the hall looked to be in fine repair, but it felt like nature needed to put her house in order. I read that the church hall floor was 'designed to accommodate dancing'.

I stood by the north porch. Somewhere to the right of it, according to the parish magazine, was a sea urchin in the wall, and I wanted to find it. I was leaning down to look at the lower flints when an elderly couple came out of the door. The lady was carrying an oasis of browning flowers. I stood up and said hello.

'Would you like to go inside?' the man asked. He was wearing a tweed jacket and was quite stooped, but he was spry nonetheless and clearly wanted to share his church.

'I'd love to, if I'm not disturbing you,' I said, and he stepped back into the porch to hold the door open for me.

It smelled of a hundred Kent churches. Dry, chalky must. A hint of paint. Damp paper. It was strangely reassuring. The walls were white except for the stonework of the windows and the chancel arch. The nave was narrow and furnished with wooden pews. It was an anti-Tardis, smaller on the inside, as the flint walls were at least half a metre thick. There was a pretty window entitled 'Love' by Morris & Co, made in 1909. The figure of Love gazed down to the right, looking wistful, dressed in a demure red gown. She looked prepared for disappointment.

There was a tower here, once, to the south, said to have collapsed following a lightning strike in the 17th century. On a stone arch thought to have led to the tower, was a single stone-carved early Norman head – 'very broad, with lizard eyes, and the mouth, broad and grinning, stuck full of teeth', according to George Clinch's *Antiquarian Jottings Relating to Bromley, Hayes, Keston, and West Wickham, in Kent (1889)*. It was well described. 'The demoniac grin and distended cheeks remind one of the grotesque heads, symbolic of the festivities connected with Whitsun ales, which in mediaeval times were carved in churches, &c.'

The Whitsun ales were a celebration. Sport, music and dance with ale brewed and sold to raise funds for the parish. There were maypoles and morris men, the whole money-making scheme feeding the later myth of a Merrie England that never was. But while there was no pagan root to the Norman festivities, I stared into that face and something much older laughed back. Older than a thousand years of devotion to a Christian god. It was a face of the earth itself.

The collection box was by the door, and I slipped a note in the slot. The lady and man were coming back in, she carrying

a fresh oasis display of red-gold chrysanthemums.

'Did you like it, dear?' asked the man.

'Very much,' I said. 'Thank you. The window is beautiful.' I gestured to the William Morris glass.

'I'm glad,' he smiled widely, with teeth that were all his own, and I fancied he glanced briefly over my shoulder to the Norman head, and back at me, still smiling.

I stepped out into the day, breathing deeply to get a millennium of enclosed space and more out of my head. I couldn't walk the church's perimeter because the hall was built up to the south fence, so I turned to the rear of the churchyard. Here were the new graves – and the children's graves, which were sad in a way no other graves can be. Most took the form of flat, engraved stones, but one in particular stood out, because it had a shoe on it. This one was not a child's grave, it commemorated a 'husband and father' whose name was lost to the grass and the shoe, but who was 'loved by everyone who knew him'. The shoe was a single canvas plimsoll, probably a child's size two, with white and yellow daisies printed on a black ground. The heel rested on a candle placed at the top right of the plaque, and inside the shoe was a bottle of pink nail varnish. I found it almost unbearably moving. In that single gesture of dedicating a shoe, almost two thousand years of history collided. History lived within a stone's throw on this green valley terrace.

I went to the church gate and dialled Arben the cab driver. He said he'd be twenty minutes.

chapter eleven

GOING HOME

Just weeks later I was back at the well. I had resolved to cross the geological divide and walk home to the chalk before the year was out, and I had days left to do it. I wanted to be a hill-fort person and walk their paths.

But 'you can't step out of your century', we were told by our metallurgist professor, in the week when we learnt to smelt tin from ore. Our culture, preconceptions and beliefs are like skin, hair and blood, and are knitted so deeply that we can imagine – but can't really truly conceive of – life outside our own time. Still, I wanted to walk it and here I was.

It was a rare midwinter morning of no cloud, and it was mild. The sun seemed so low, it shone barely above my shoulder.

The well was busy. A bunch of flowers – fancy striped daffodils still in their cellophane – was lying half in, half out of the pool. A biro note said: '*Dear Mum and Dad, love you always.*' A girl's name was written below, followed by '*and the dog*'.

Kids in Wellington boots wandered in and out of the

pool, mesmerised by their own ripples. One little girl had waded too deep and was explaining to her parents through a series of sounds and points that her boots were full of water. There were family dogs. And more parents with small kids. Beyond the well and the trees, the top pond glinted a faded silvery blue. I took a picture. It looked like childhood, an iPhone recreation of a Super 8 memory.

I climbed the steps out of the hollow, now extremely slippery with mud. I was grateful for the wooden rail. At the top a robin was waiting on one of the rail posts. He was big and bold, and I suspected he owned the car park, or at least ran that corner of it. I dug about in my rucksack for nuts, but by the time I'd crushed them sufficiently to make them snack sized, he'd gone.

I left the car park and crossed the road to the footpath that led up and away from the well. It was marked by a sign that pointed to a gently climbing path through woodland. The sign read 'Downe 1½ miles via Wilberforce Oak.' And so I began, off the Blackheath Beds with its billions of small pebbles and home to the flint and chalk.

There was not much geology visible through the mulch of black leaves and mud, but I could feel the faint crunch of pebble and sand with each footfall. It was less a sound, more a sensation, like eating a sugar sandwich.

To the left of the path was a substantial five-foot post-and-wire fence enclosing a tangle of wild wood that belonged to Holwood House. An identical unfenced tangle lay to the right. 'Private Property' signs were fixed to the fence every hundred metres or so, though you'd have had to go some to vault it, and if you did, you'd find yourself in a muss of unchecked trees, ivy and thorn. Somewhere beyond and above was the hillfort and its western gate. I doubted I'd even get a glimpse.

The sun was sidelighting the trees to my right, and the ivy leaves reflected it a thousandfold in silver. The scattered light made it hard to see fully, even with sunglasses on. As I walked, the landscape to my right fell away until the path was hugging the side of a steep hill. There was a clearing up ahead. I was at the top – or as far to the top as the path, still with its fence to the left, would allow.

I had never seen the Wilberforce Oak, and while the old tree had all but gone, its companion didn't disappoint. The second tree stood a short way from the hilltop in a grassy clearing, on the downslope to a wooded valley. The sun was low and bright in the open, and I could just about squint to a distant ridge of trees. The entire landscape seemed little more than a stage for the sibling oak, which, though technically dead, dominated the sward like the god it clearly was. Its girth was well over a metre-and-a-half; the nubs of former branches had taken animal form, except for one, which jabbed skyward like a horn. The tree had a hundred faces in my imagination: its main one roared into the sun with a voice I was too mortal to hear. On its forehead lounged a young bear, worn like a giant fascinator. Walking widdershins, I saw a lizard, a deer, and a likeness of Herne himself. On the shadow side, I peered through a hole into the hollow heart of the tree where the sun, blazing through its open mouth, turned the inner oak to beaten gold.

By the side of the path above me was a wooden seat for walkers to appreciate the view. Behind, and through the wire fence of Holwood, a metal notice mounted on a wooden panel was engraved with the following:

> From William Wilberforce's diary, 1788.
> *At length, I well remember after a conversation with Mr Pitt in the open air at the root of an old oak at Holwood, just above the steep descent into the vale of Keston, I resolved to give notice on a fit occasion in the House of Commons of my intention to bring forward the abolition of the slave trade.*

I sat on the bench and screwed my eyes into the afternoon light and the view. The wide crumbling core of the original oak had been scattered with deep-pink carnations. I liked the flowers and the fact that someone had thought to bring them. A young oak was growing from the old tree's bones. I liked that, too. It was hopeful.

I poured some coffee from my flask and ate three mini mince pies straight from the packet in just a few bites. I thought about the repercussions of the events discussed right at this spot and the millions of lives affected. I wondered if the people of the hillfort had slaves. Probably. Farming in the early days is now considered impossible without. Maybe the slaves came from another tribe or another place. Maybe they were particular families, or women. If not people, the farmers had animals. What are domesticated animals if not slaves?

I took a selfie in the sunshine, sitting on the top of the bench by the sign. The tree god's big, angry, reindeer shadow face stared at me. I stared back. I hadn't the nerve to take a selfie with it. Before I set off, though, I walked down to the tree and touched it with my fingertips. It was warm. It glared.

I thought of the bear in the tree as I took the path again. And of bears generally. Were there brown bears here when the hillfort was occupied? How would it be to walk these paths knowing you risked attack, not just by brigands but by a creature that might eat you? It's thought that there were few bears in Britain by the Iron Age; the only way of being sure is by finding their antique bones in a secure, datable context, such as a cave, and there's a lack of those on the Downs. Bear numbers in England rose during the Roman occupation, when they are supposed to have been imported from the continent for entertainment. It's argued that they survived into medieval times, but, if they did, they would have been chased to the wild fringes of the landscape. No, after all that, I didn't think the hillfort builders were much troubled by bears.

Wolves were a different thing. They survived into the early years of the 16th century in the south, and on downland would have preyed on deer and sheep. They could also root out a corpse from a grave, according to both lore and record, and I wondered if this in part prompted an Iron Age shift to cremation. (Somewhere close by, probably under today's ploughsoil, or maybe on the wild commons, lay the ringed postholes of the hillfort builders' roundhouses, and almost certainly their dead, burnt bones with twisted treasures in rough, handsome urns. They have not been found. It was almost too large a thought for the day.)

In the end, the wolves were trapped and hunted to

extinction as their forests were cleared, as were the wild boars, but in 1987 the boars had a reversal of fortune when a tree fell on the enclosure of a small captive colony in West Kent during the Great Storm. The escapees prospered, one family periodically alarming drivers on the A21 near Tunbridge Wells.

I have never seen a boar. I don't think they've made it this far east. Though I often see deer on the Downs. One used to watch me some summers back as I walked the paths of Chalky Luggitt. She would hide in the barley, navigating the tramlines, and track me (or so I chose to believe), her little head rising from the crop in unexpected places like a periscope, before dipping again. The deer spend their days in the woods near where the unfortunate Roman got an axe in the head, spooking walkers or being spooked by dogs, and eating flowers from the graves in the churchyard, leaving the squirrels to be blamed.

The descent was steeper than the climb. The path crossed the long, brick-block driveway to Holwood House, and I saw nothing through the metal gates but more drive and more woods. The path dipped dramatically after the drive, becoming a skiddish hollow way of leaves, pebbles and grit. And then, under the leaves, there was a streak of white about the size of a forearm. Then more white, and a piece of weathered blue flint half out of the leaves to the path-side. Pebbles still spilled over the chalk bedrock, and continued to do so, almost to the valley floor. But I was geologically home.

I think I expected something more. Perhaps I'd made too much of the idea. I hadn't felt like a hillfort person, though maybe there were too many distractions, or I hadn't tried hard enough. And there are certainly places I walk where the delineation of geology is sharper. All the graves in the churchyard of St Giles the Abbot in Farnborough, as an example, are cut deep into the chalk, with great dumps of the spoil thrown into the gravel cuttings in the woods below. Just metres to the east, John Lubbock and his descendants are buried in greensand. The line is clear. You step over it. I walked through the family plot one spring, as a cheerful man dug the grave for Eric Lubbock, Liberal peer, 4th Baron Avebury and John's grandson. The man let me watch as he carved out the orange sand with a digger. It was easier than the chalk, he said. A couple of years later, sarsens appeared, one at the head of Eric's grave, and a larger one on the central mausoleum. It was a beautiful tribute. The sarsens came from Avebury in Wiltshire; the same dense boulder-shaped sandstone as was used to build the Avebury monument. Erratics, they would be called now. Stones that don't belong in the place they are found.

At the bottom of the path was a crossroads. All of them lanes and all busy with cars. I took the footpath east that followed the valley floor. Robins in the hawthorn hedges busked as I walked, and I left them some nuts on a post. Looking north across the field beside me up to the woodland I had just walked, I got my first view of Holwood House. Two cedars of Lebanon stood like stage curtains, framing a long, golden-yellow, two-storey Palladian villa. It was beautiful. And this low, tight, muddy path was the only place any member of the public could see it. It was not William Pitt's house I was looking upon; that had been pulled down after the sale of the house to a Mr John Ward and rebuilt in 1825 by the architect Decimus Burton.

Private land lay ahead, and the path directed me across to the other side of the lane to a tarmacked hollow way that rose steeply to the west end of Bogey Lane – and familiar territory.

The sun shone warm on my face and I felt safe, I reminded myself. Happy as a duck on water or a spider in its web, walking as I was on a landscape I considered my own. There were no bears. No wolves to beware of. No poisonous snakes. No highwaymen or highwaywomen. All chased from the Downs centuries before. All gone except for my ridiculous

fear of a place with a name that doubtless meant nothing. I posed for a selfie by the road sign, then immediately posted it on Facebook, hoping that childishness would throw a fire blanket on my discomfort.

Not scared, I stepped onto the dirt-and-flint lane.

Some thirty paces along, a large tree had fallen downslope into the neighbouring field. Its roots had left a hollow scoop in the ground and a large vertical wall of ginger clay and flint. I examined the roots, hopeful of a Saxon hoard, or at least a coin, though that was wishful thinking as the tree was at best two centuries old. Not the track, though. The track was ancient, and you never know.

The flint more than made up for the lack of gold. All kinds were caught in the root-and-clay matrix. Small, frost-shattered nodules partially cortexed, a swash of pebbles, small white-patinated natural shards, all things that looked promising as something besides naturally modified flint.

I paid attention to a few caramel-coloured pieces in case they were fragments of some greater Palaeolithic tool, but they were all naturally shattered – at least, as far as I could tell.

The winter rain had made everything wet, and, to the side of the tree, spectacular mushrooms were growing on some older wood. I undid my rucksack and pulled out a small camera that was good for macro. It was a whole world of penicillin blue in slender antler form. I put the camera in my coat pocket in the hope of seeing more fungi later.

I'd not seen many people on the walk. Wrong time of year, wrong time of day. Wrong leaves on the path. But someone was coming along the lane towards me. A man with a shaven head was gripping the lead of a huge black dog tight to the collar. It was more like the coal-coated, red-eyed Black Shuck of English legend than domestic pet. The dog was panting and straining against the man's hold. The tension made the two of them vibrate.

He spoke, urgently. A growl: 'Aaow big is it?'

He was looking both at me and beyond me. Almost nervously. I didn't know what I was supposed to say.

'How big is what?' I asked. His shuck had fire in his eyes.

'Yer dog!' I was clearly stupid. 'Aow big?'

'I don't have a dog...?' I said. He was closer now.

'You got a lead!' he said accusingly, nodding with his head to my pocket, where the strap of my camera looped out.

Ahh, I clearly needed to prove my lack of dog to him. I pulled on the strap and the camera came out.

'It's a camera.'

'Uhh. No dog?'

He was almost level with me, and his shuck looked like he could swallow a corgi whole.

'No. I work in London. I don't have time... No dog.'

'Uh,' he grunted.

I stepped into the hedge and the pair scuffled past in a fug of unfocused energy and aggression. I turned to look after them. The shuck was trying to turn his head back but his neck was too thick.

I was utterly unnerved. Scared physically and shaken psychically. And I didn't owe anyone an explanation.

'I have a dog!' I wanted to shout. 'My dog is as big as a field. As big as all the Downs and as heavy as all the flint and chalk underneath them. My dog is made of clouds. She's everywhere! It's just that you can't see her, and I don't like pretending she's not there.'

A few metres farther down the lane was a gap in the hedge. I stepped through it and into the field. I was facing north, looking down to the road that had once been water. The Field of the Ghost River, its shining twin, rose on the opposite side of the road from the valley floor to the wooded ridge. It basked in the afternoon sunshine. I wanted to get there. Put distance between myself and the shuck wrangler.

This was the sterile field, anyway. Sterile in terms of human habitation. I'd walked it a hundred times and never found more than a fossil. I hurried down the path and crossed the road to the north side.

Three paces into the Field of the Ghost River, in the sharp, faded stubble, I found a large flint knife with a slight curve, broken at the base. It was, like most of the flint on the Downs, shiny white with thick patination.

Two steps farther on I found an oval scraper with a serrated edge. Then a core. I looked around to see if John Lubbock and Van were up ahead. I was finding too much. There was a dog walker on the ridge by the trees, but no ghosts. Then I saw the axe. Smallish, about the size of an old Nokia phone, it was lying half in, half out of the clay, knapped and patinated surface uppermost. It had a flat butt and pointed tip. I reached down and turned it over. Its ventral surface was almost flat and had some kind of concretion on it. It was not beautiful, but it was old. Maybe even 9,000 years old. It would have been used not by a farmer but by someone who lived by hunting and gathering. The people who used this would have known bears.

I took a photo of it in its location. Then of me with it. I managed to get a high-flying crow in the picture and was pleased.

I followed the plough line east. These were almost perfect fieldwalking conditions. A long season of weathering had washed the visible flint completely clean, making it easy to see, all but for the shadow striations of the stubble putting stripes onto everything.

So then I saw a fossil. A lovely broken micraster about six centimetres across. It was an internal cast with beautiful moulding in its thick white rind. I held it up to the west to photograph, against the distant ridge of the hillfort. Time across time.

I walked on. Small brown birds rose, heckling and sparkling in the dipping sun. I was nearly at the edge now and was almost relieved to be leaving the field, which had become some kind of god of abundance. I felt almost embarrassed at my finds.

I stepped over the small rut that ran parallel with the path, glancing down as I did, and saw a lovely nodule. It seemed to have all of its cortex. I bent down and pulled it out. It looked like a flat chicken. It was whole and heavy. I brushed off the clay with my fingers. My neighbour kept hens and had a herb garden, and I thought she would love it. I weighed up the idea of carrying it home, but it was big and I decided against it.

Instead, I stashed the chicken flint at the base of a tree in a small copse – one I often use – and walked down to the bridleway. It was dark on the path now, and I didn't meet a soul. I came level with the gatehouse of John Lubbock's estate, a small, flint one-storey cottage just visible through the trees on the far side of the road. It was called – with cheering lack of affectation – Flint Lodge. Lights were already on in the windows.

I left the woods and wriggled through the fence into

Chalky Luggitt. Green shoots were driving through the sticky valley-bottom clay, so I stuck to the perimeter path, glancing at the flint in the fields. I saw what I thought was a core, but it wasn't.

At the midway gate I crossed the road to Luggitt's southern twin and followed the path up alongside the trees. The sun was low over John Lubbock's estate and the sky to the west was orange fading to peach. The crows had gone to roost, and I wondered if I might see – or at least hear – an owl.

It took two minutes to cross the darkening woods on the ridge, where one of the last gibbets in England was said to have stood, though some said it was near the site of the allotments in the corner of Chalky Luggitt, with the old turnpike road apparently doglegging around the field because of the smell of the corpses. A gibbet was an iron cage – large enough to contain a man – that swung from a high wooden post. An expensive piece of theatre, gibbeting was used for the public execution and display of felons such as highwaymen and smugglers, with the aim of deterring the populace from criminal acts. Both places fitted the bill for classic gibbet location (hilltop, outside London, major route, in this case to the coast), but I've been assured it was indeed on this hilltop crossroads that the gibbet stood, clear, in those days, of the thick woods I'd just traversed.

I didn't take path down to the village. The light was glorious and I wanted to see my house on the next hill before I went home. I ducked through the fence into a grassy field – a

squarish plateau with woods on three sides. It was one of the highest points for miles. Hawthorn fringed the lynchet to the south, from where two further fields sloped away to the valley floor and the A21. From a wide gap in the hawthorns, I could see a great chalkland vista where the Downs creased and rose to a series of wooded ridges that blended into one on the horizon. All the fields and the few lanes and houses were fading in the night rise, but the ridge stood distinct against the light.

I could name the woods: Rounds Wood, Charm Wood, Hook Wood, Birthday Wood... And below, the oblique kite shape of Dead Car Wood. I wondered what the seasons had done to the cars' remains.

Through the trees to my right was a deepening orange glow. The sun had gone. To my left, in the sky above the hill of my home, was the moon, quite high now and last quarter. It hung above the Belt of Venus – a sugared-almond-pink arc, turning deeper by the minute. The glow filled a third of the sky in a band above a low arc of pale cerulean blue. The sun, having dipped below the western horizon, was blasting Earth's own blue shadow back up onto the atmosphere in the east. It was simple physics and pure magic, dependent on nothing more than a clear sky and the right atmospheric conditions.

This evening's anti-twilight arch (a clumsy name for such an elegant phenomenon) was of an intensity that rendered the whole world pink and blue at the same time. I wanted to lie in the field, right there on the wet grass with my flint-filled bag as a pillow, axe in one hand, fossil in the other, and breathe in the whole big wonder of it until I had sucked the whole world and everything on it and above it inside of me. Then I would breathe it all out and everything would be right back where I started.

I sat down on the grass near the hawthorn gap. A plane drifted overhead, pink neon; still day for the passengers on board. A small flock of jackdaws flew west. I stayed sitting. I didn't lie down and look at the sky. I didn't want a dog walker to come by and think I was dead.

The blue arch grew. And so did the pink, reaching right up over my head. The hawthorns edging the fields below were now a bruised plum, the dead grass strands along the green field's edges a straw pink.

Over on the far Downs, the odd car with headlights dipped under the line of tiny distant pylons with their gossamer wires. Nothing else moved. I could hear the faint hum of traffic and life. I wished I could put everything on pause right there. Everything perfect.

I became aware of a dog, though I'd not been conscious of his approach. He stood silently, a little way to my right. Dogs know fear and I tried not to be afraid, but I was wary. I needn't have been.

He walked closer and sat next to me, looking, as I was, across the valley. I had planned to sit there for as long as I had light to see, but after a few minutes the dog got up and was clearly waiting for me to do the same.

So I did. I followed him to the edge of the hawthorn gap and stepped carefully down the animal track of the steep lynchet. This was the place of the big flint cairn, and, in the twilight, I could see that it had been rebuilt, very recently, by the look of it, and well. The old skyward-pointing nodule was surrounded by a supporting coronet of smaller pieces, creating a conical, cake-like structure. It looked like a giant, pointy meringue or a pièce montée.

The dog turned west and headed along the lynchet in the direction of the woods. There was no pink left now in the sky, but the blue scatter – and the moon – was plenty to see by. The lynchet curved, and, as I rounded the bend, I saw a man. He was around fifty metres from me, facing the woods, one foot on the upslope of the lynchet, the other – to balance – on the field. He was rangy, with rough, tweedy trousers and leather walking boots. Beside him, in the field, was his coat, and a large pile of flint nodules. He was stacking them carefully on a flat part of the slope, making a second cairn.

He was oblivious to our approach, and I watched quietly for some moments as the cairn took shape. He had a beard.

Then the black poodle beside me leant softly against my leg. I scratched Van's warm ear softly with a finger and he gave a low woof.

John Lubbock turned round.

GLOSSARY

Anglian stage The British name for the most severe ice age in the past two million years, beginning around 478,000 years ago and ending around 424,000 years ago.

Astraphobia Extreme fear of lightning and thunder, from the Greek word *astrape*, meaning lightning. See also *Brontophobia*.

Aurochs Species of cattle and the large ancestor of modern domestic breeds, probably extinct in Britain around 1000 BC. The last aurochs of a small herd died of natural causes in the Jaktorowska forest to the south-west of Warsaw, Poland, in the early 17th century.

Biogenic A substance produced by, or made from, living organisms.

Braided river A river with multiple channels that separate and merge, with gravel banks or sand bars in between, creating a braided pattern across the landscape when seen from above. The channels shift over time, forming, absorbing and re-forming the bars.

Brontophobia Extreme fear of thunder, from the Greek word *bronte*, meaning thunder.

Brontoscopic calendar Divinatory calendar of the Etruscans – people of pre-Roman Italy –whose deities revealed themselves in natural phenomena such as bird flight or lightning. The calendar lists every day of the year, and the possible outcomes of thunder on that day. It was said to be the work of the Etruscan prophet Tages around 680 BC, transcribed by the Roman scholar Nigidius Figulus and later translated into Greek by the 1st century Byzantine antiquarian John Lydus. Etruia was a region of Central Italy that comprises modern Tuscany and part of Lazio and Umbria.

Bulb of percussion The bulge on the ventral (front) surface of a flint flake, directly below the point of impact that caused it to separate from a nodule (or core). The flake bears the positive – convex – bulb, the nodule bears the negative – concave – scar of detachment. Bulbs can be pronounced or subtle, depending on the force of the strike and whether the hammer was hard (made of stone) or soft (antler).

Bulbar scar A small chip on the bulb of percussion caused by the impact of the hammer strike.

Chatter marks Small crescentic (crescent-shaped) scars found on pebbles. These are the marks of violent motion as the flint is thrown together in turbulent waters, tumbled and worn smooth over a period of several millennia.

Chert A microcrystalline variety of quartz, mineralogically the same as flint, but usually pale buff to brown in colour and granular in texture, commonly found in limestone deposits. Sometimes chert is used as the umbrella term for both flint and chert, with flint considered to be a darker variety of the rock. Not all geologists agree on this, but they do agree that flint is the name for the nodules found in chalk.

Cleavage A rock or mineral's natural tendency to cleave – or break – along a particular plane, called a plane of separation, according to the arrangement of its atoms. Not all minerals have cleavage. Quartz, opal and flint have none, and their irregular breakage is called fracture.

Cretaceous The geological period from around 145 to 66 million years ago. The name comes from the Latin word *creta*, meaning chalk, as the chalk formed in this period in the many warm, shallow seas that covered the planet. The mass extinction at the end of the Cretaceous period, thought to be caused by an asteroid strike, saw the end of 75 per cent of life on earth, including the dinosaurs.

Coccolithophore A single-celled plant-like organism (a plant plankton, also called phytoplankton) that lives in the warm upper layers of the ocean. Coccolithophores protect themselves by building coccoliths – microscopic plates of calcite – around themselves, with a single organism might having as many as 30 scales. The organisms exist in huge numbers and, when they die, the calcite from their scales is dumped back into the ocean. In the

Cretaceous period, the coccoliths provided the material for the chalk that formed in the shallow seas.

Conchoidal fracture The smooth, curved break of a rock or mineral that looks similar to the rippling growth lines on a clam or mussel shell. Flint, obsidian and glass all have conchoidal fracture.

Core A piece of flint from which flakes or blades have been removed.

Cortex The natural 'skin' of flint, made from powdery opaline silica. White when fresh from the chalk, the colour ranges from buff to black, as chemicals from soil or water tint the porous rind over time.

Crinoid A marine animal also known as a sea lily, because of its habit of attaching itself to the seabed by its stalk in its juvenile form. Its family includes star fish and sea urchins.

Debitage The waste flakes and chips resulting from flint knapping.

Dene hole A vertical shaft dug into chalk with a number of cave-like chambers extending out from the base. They are chalk mines, dug to access fresh chalk for use as fertiliser. The chalk provided extra calcium for clay or acid soils, and improved drainage if the soils were particularly heavy. Dene holes were known in Britain from Roman times and were still being dug in the early 20th century.

Earthwork Works in or on the ground that show past human activity. They can be positive, such as barrows, mounds or banks of earth, or negative, such as moats, ditches or tracks.

Echinoderms A group of marine animals. They include Echinoids (see below), Asteroids (starfish), Holothurians (sea cucumbers), Crinoids (sea lillies and feather stars) and Ophiuroids (brittle stars).

Echinoids Known by their common name of sea urchins, this group of spiny marine animals consists of around 1,000 species today, living on ocean floors across the world. They evolved around 450 million years ago, and were present in the Cretaceous period, living – and dying – on the chalky seabeds where their bodies sometimes became fossilised in the silica-rich waters, eventually becoming cast in flint.

Erratic A rock that appears in a different place from its original geological location. It might have been shifted by glaciation, transported by a river, or carried by a human or other creature.

Fairy loaves The name given in folklore to fossilised sea urchins because of their resemblance to miniature loaves of bread. They were believed to be lucky.

Fieldwalking Walking fields in a systematic manner with the aim of counting the number of – and sometimes collecting – artefacts on the surface, to assess possible human or hominin activity in the past.

Foraminifera Marine micro-organisms, many of which have calcite shells which were a major contributor to the formation of chalk in the Cretaceous period.

Fracture The irregular breakage of a rock or mineral.

Greenstone A greenish-grey volcanic stone valued in prehistory for its colour and durability for tools.

Hag stone A stone with a naturally formed hole, usually flint, believed to have magical properties. Also known as an adder stone.

Hammerstone A stone used to strike flakes from a flint nodule or core, and to fashion the flakes into tools.

Hillfort A hilltop fortress of the Late Bronze Age or Iron Age. Once thought to be defensive, the purpose of these sites may vary and is not fully understood.

Hinge fracture A rounded end to a flint flake that looks like a cup rim and is the result of the shock waves doubling back on the flake, rather than rippling down to the end. It is caused by a fault in the flint or a mis-strike with the hammer.

Hollow way A track worn to mud and stone by human and animal traffic and hollowed deep in the middle as the loosened earth is carried downslope by rain.

Hominin A member of the tribe Hominini that includes humans and close extinct relatives.

Homo heidelbergensis An early species of human. *H. heidelbergensis* made tools including stone tools and wooden spears, made fire and hunted large game. The species is believed to have lived 1.3 million to 200,000 years ago and was present in England from around 500,000 years ago.

Hoxnian stage An interglacial (see below) that followed the ice age known as the Anglian stage. This warm phase began around 424,000 years ago and lasted around 50,000 years.

Interglacial The warm phases that alternate with the ice ages. The current warm phase, known as the Holocene, began around 12,000 years ago and corresponds with the Mesolithic, or Middle Stone Age, in most of Europe.

Knap The deliberate shaping of flint by a series of blows with a hard or soft hammer to remove flakes and make the raw nodule into specifically shaped tools.

Loess Sediment deposited by the wind, made of clay, sand, silt and small quantities of organic dust.

Lynchet (also known as strip lynchet) A feature of the old field systems of the British Isles that takes the form of an earth terrace on a hillside. Formed by ploughing, as the plough loosened the soil of a sloping field and caused it to slide to the downside of the field over time, making a bank that curved around the contours of the hill. On some steep hillsides lynchets were humanly made.

Mesolithic The Middle Stone Age, a period from about 11,000 to 6,000 years ago (9000 to 4000 BC) in Britain. Once thought to be exclusively nomadic, its people are now believed to have occupied seasonal camps, or even permanent settlements.

Nailbourne Kentish name for a seasonal stream. Fissures in the chalk act as springs when the water table is high, creating streams that run in the normally dry landscape until the water table subsides. Known also in Kentish folklore as Woe-Waters, they were said to be harbingers of doom.

Neolithic The New Stone Age, a period from about 6,000 to 4,000 years ago (4000 to 2000 BC) in Britain, characterised by stone and earth monument building, farming, a distinctive stone tool technology, pottery making and funerary rituals. Recent studies suggest these early farmers were migrants from continental Europe.

Opaline silica A friable and water-rich form of silica that forms the white cortex of flint. This cortex is often worn away by water on flints found by rivers or on beaches, leaving the dark flint interior exposed and smoothed.

Opus signinum A form of concrete used in Roman times made from waterproof lime and crushed and sometimes powdered terracotta pots and tile. It was a common flooring in many Roman houses and was used for floors and walls in bath houses and other buildings that required a waterproof surface.

Palaeolithic The Old Stone Age, a period from 3.3 million years ago to around 11,000 years ago, characterised by the use of stone tools. The Palaeolithic in Britain dates from around 900,000 years ago to 11,000 years ago, when Britain was occupied first by Hominins, then *Homo sapiens.*

Paleogene The period that followed the Cretaceous period, from 66 million years ago to 23 million years ago. The supercontinents Pangaea and Gondwana had already broken up in the Cretaceous, and the new continents of the Paleogene period continued their move closer to the positions they occupy today.

Paramoudra flints Also known as pot stones, these are large nodules with a hollow or chalk-filled centre, found across north-west Europe. They are thought to be the fossilised remains of sponges or the cast of an animal burrow.

Periglacial environment A landscape that is subject to cycles of freezing and thawing, generally on the edge of a region covered with ice.

Plastic virtue A geological theory that stemmed from the beliefs of the ancient Greek philosophers and still subscribed to in the 16th century. Early geologists, seeking an explanation for fossils, believed in the plastic (malleable) virtue of the earth, in which it was believed that the earth itself, driven by the cosmos, could fashion inorganic rock into organic matter. Fossils were considered to be life that had 'stalled', or life simply 'echoed' in the mineral world.

Quadrifrons A Roman arch with four facades. The Arch of Janus is the only such surviving arch in Rome; another example is the Arch of Septimus Severus at Leptis Magna, Libya.

Splintery One of the ways a mineral breaks, characterised by sharp elongated points.

Striking platform The flat or slightly slanting surface prepared by knapping on a flint nodule or core. This platform is struck by a hammerstone to remove a flake of predictable shape or size.

Tegulae Terracotta tiles, used together with imbrices, as the roof tile system of ancient Greece and Rome. Tegulae have a raised border on either side to channel rainwater down the roof. Curved imbrices were lain over the join for additional rainproofing.

Stratigraphy The layers of rock, soil and sediment that accumulate on the earth over time. The study and interpretation of these layers, based on the oldest at the bottom and the youngest at the top, has led to an understanding of timescales of the planet.

Thunderstone A stone handaxe, arrowhead, flint tool or fossil used as protection against thunder. The belief in thunderstones and their power was widespread in the Greek and Roman world and persisted in folklore around the world into the 20th century.

Tower tombs Tall masonry tombs built in a variety of heights and styles across the Roman world to house the dead. Most surviving tower tombs are found in North Africa, Syria, Lebanon and Iraq.

Tramlines The lines in a crop-sewn field where seed is not planted. The lines allow space to for crop sprayers and other farm machinery to run.

FURTHER READING

A Place in the Country: High Elms, Downe, Kent
by Ken Wilson, London Borough of Bromley Recreation Department, 1982

The Archaeology of Kent to AD 800
edited by John H. Williams, Boydell Press, 2007

Flint: The Versatile Stone
by H.J. Mason, Providence Press, 2000

The Kent Downs
by Dan Tuson, Tempus, 2007

Making a Handaxe
by Mark Edmonds, Group 6 Press, 2022

The Man Behind the Bayeux Tapestry
by Trevor Rowley, The History Press, 2013

Myth and Geology
by L. Piccardi and W.B. Masse, Geological Society of London, 2007

The Nature and Subsequent Uses of Flint
by John W. Lord, 1993

Prehistoric Flintwork
by Chris Butler, The History Press, 2005

The Roman Villa Site at Keston, Kent, First Report (Excavations 1968-1978)
by Brian Philp, Keith Parfitt, John Wilson, Mike Dutto and Wendy Williams, Kent Archaeological Unit, 1991

The Roman Villa Site at Keston, Kent, Second Report (Excavations 1967 and 1979-90)
by Brian Philp, Keith Parfitt, John Wilson and Wendy Williams, Kent Archaeological Unit, 1999

Shipwrecks of the Goodwin Sands
by Richard and Bridget Larn, Meresborough Books, 1995

Some Notes About Prehistoric Flintwork
by Paul Hart, Trust for Thanet Archaeology, 2021

The Star-Crossed Stone
by Kenneth J. McNamara, The University of Chicago Press, 2011

ACKNOWLEDGEMENTS

I have read many books, monographs and research papers, visited many museums, attended lectures and spoken to several knowledgeable people in the course of finding out about flint. Any misreadings and misinterpretations are mine.

Andy Martin, Anne Mitchell, Patricia Monahan, Christina Rodenbeck and Guy Sawtell read the chapters of the manuscript as I wrote them and were phenomenally encouraging.

And I'll always be grateful to Tim Reynolds for looking at my finds and sorting the humanly struck from the natural in The Lamb, Bloomsbury.

The following people read part, or all, of the manuscript, offered information on flint and other aspects of history and prehistory, gave me lifts, did sums and tracked down esoteric information. Some just shared my enthusiasm for flint and for that I'm extremely grateful:

John Ainsworth, Alice Amabilino, Laura Basell, Frank Beresford, Dan Bourne, Glenda Bourne, Isabella Bourne, Nick Card, Juliet Clarke, Paul Craddock, Glenn Dakin, Joe Dyson-Hawkes, Emily Dyson-Hawkes, Mark Edmonds, Katy Everett, Stephen Fall, Simon Gray, Christopher Gregory, Curtis Hawtin, Lewis Hawtin, Simon Hawtin, Robert Kimber, James King, Claire Lister, Matt McAllister, Jo McKenzie, Iain McKenzie, Thelma Morley, Alice Peebles, Stuart

Rathbone, Ben Robinson, Ben Saunders, Steve Scanlan, Paul Southcombe, Anne Teather, Sigurd Towrie, Karen Wallis, Alison Wilcock and Tim Wilcock.

Thanks, too, to Dan Hiscocks, Simon Edge, Nell Wood and the team at Eye Books for their help and enthusiasm in bringing it into being.

Finally, thanks to local farmers the Wilson family for permitting the public to walk the tracks around and across their fields. Right across the Downs we find flint by the farmers' grace, so please respect their lands and never step off the tracks when the fields are seeded or crops are growing.

also from Eye Books

LOCAL

ALASTAIR HUMPHREYS

A search for nearby nature and wildness

After years of expeditions all over the world, adventurer Alastair Humphreys spends a year exploring the detailed local map around his home.

Can this unassuming landscape, marked by the glow of city lights and the hum of busy roads, hold any surprises for the world traveller or satisfy his wanderlust? Could a single map provide a lifetime of exploration?

Discovering more about the natural world than in all his years in remote environments, he learns the value of truly getting to know his neighbourhood.

An ode to slowing down, Local is a celebration of curiosity and time spent outdoors, as well as a rallying cry to protect the wild places on our doorstep.

Agile, wryly funny and wise - **Robert Macfarlane**

A paean to the benefits of determined noticing. What really shines through its pages is Humphreys' omnivorous curiosity - **Financial Times**

Thanks to some genuinely thoughtful writing about planet, place and political purpose, Humphreys finds beauty in the scruffy margins and makes readers look anew at what might easily be familiar or forgotten - **The Observer**

Alastair Humphreys is the consummate roamer: big of heart, curious of mind, light of step - **Amy-Jane Beer**

If you have enjoyed *Flint*, do please help us spread the word – by putting a review online; by posting something on social media; or in the old-fashioned way by simply telling your friends or family about it.

Book publishing is a very competitive business these days, in a saturated market, and small independent publishers such as ourselves are often crowded out by the big houses. Support from readers like you can make all the difference to a book's success.

Many thanks.

Dan Hiscocks
Publisher, Eye Books